HISTORY OF TECHNOLOGY

Thus, considering a typical doping concentration at the source end of about 10^{17} cm^{-3} and about 10^{15} cm^{-3} at the drain end (epitaxial layer doping), the minimum channel length required to avoid punch through is about 2μm for a 200V device. It should be noted at this point that since the doping concentration at the metallurgical junction does not change instantaneously, the peak is not located at the source edge but is slightly shifted into the channel. Therefore, the minimum channel length calculated above must be increased by the same amount as the distance required to reach the peak (0.3 to 0.5μm). Moreover, it should be noted that the doping concentration profile in the channel controls the threshold voltage of the device as well. Since a minimum amount of boron doping must be present to avoid punchthrough between drain and source, it sets a minimum value of the threshold voltage. The controllability of the latter as a function of the doping levels can be found in [10].

4) The lightly doped drain extension length Ldd and the field-plate lengths L_{g1} and L_{g2}: As was mentioned in the previous paragraphs, the presence of a lightly doped drain extension allows lateral devices to withstand the voltage drop between the high-voltage drain contact and the channel. Nevertheless, the simple extension of the drain length L_{dd} is not sufficient to overcome the high-voltage constraints. In fact, because of surface effects (chapter 2 section 2.2.2), avalanche breakdown would occur at the surface of the drain-to-channel junction without the presence of a gate field-plate, whose utility was already discussed in the same section. Therefore, the gate must extend over the thin oxide layer by L_{g1} and then it must step over the thick oxide layer by L_{g2}. Hence, the depletion region of the drain-to-channel junction terminates under thick oxide. It is important that the gate overlap stops short of the drain n^+ contact, otherwise avalanching of the surface near the n^+ contact degrades the voltage capability of the device [11]. Because of the two dimensional distribution of the electric field and potential in the depletion region located under the field-plate, the determination of the lightly doped drain length L_{dd} and field-plate lengths L_{g1} and L_{g2} is a complex task. Numerical methods are usually used to solve Poisson's equation in this region. However, first order considerations can lead to a rough estimation of these parameters. Indeed, the drain-to-channel junction with a simple field-plate of minimum length L_g, is shown in Figure 3.6. The optimal field-plate length is considered when, at maximum reverse drain voltage, the cylindrical portion of the depletion layer created by the field-plate, intersects the one of the junction curvature. In this way, the electric field lines at the surface are spread along the resulting depletion layer, as shown by the hatched area in Figure 3.6. Note that in the previous consideration a cylindrical approximation was made on the depletion layer at the edge of the field-plate. Hence, the maximum efficiency of the field-plate effect is obtained when the field-plate length L_g satisfies the following Pythagoras law

$$L = r_j + L_g = \sqrt{r^2 - X_d^2} \tag{3.12}$$

Replacing r by r_j+W, the minimum field-plate length can be expressed as

History of Technology

Volume Eighteen, 1996

Edited by
Graham Hollister-Short and
Frank A.J.L. James

MANSELL

First published 1997 by
Mansell Publishing Limited, *A Cassell imprint*
Wellington House, 125 Strand, London WC2R 0BB, England
PO Box 605, Herndon, VA 20172

British Library Cataloguing-in-Publication Data

History of Technology – 18th volume (1996)
1. Technology – History – Periodicals

ISBN 0-7201-2345-3
ISSN 0307-5451

Typeset by York House Typographic Ltd, London
Printed and bound in Great Britain by Biddles Ltd, Guildford and King's Lynn

Contents

Editorial Note vi

The Contributors vii

Notes for Contributors viii

A. RUPERT HALL

Theory and Responsibility in Science and Technology 1

M. T. WRIGHT

On the Lift Pump 13

IAN INKSTER

Discoveries, Inventions and Industrial Revolutions: On the Varying Contributions of Technologies and Institutions from an International Historical Perspective 39

R. L. HILLS

James Watt, Mechanical Engineer 59

ALAN L. LOUGHEED

Technological Advance in the Manufacture of Chemicals: The Case of Cyanide, 1888–1930 81

JOHN K. BRADLEY

Putting the Wind up the Pilot: Cloud Flying with Early Aircraft Instruments 95

STEPHEN N. TRAVIS

'Seeing' Is Believing: The Development of Microwave Radar in Britain, Summer 1940 113

TERRY GOURVISH

Diffusion of Brewing Technology since 1900: Change and the Consumer 139

Contents of Former Volumes 149

Editorial Note

Dr Frank James has ceased to be co-editor of *History of Technology*, with effect from the publication of Volume 18. The Publishers thank him for his contribution to the journal, and also the Royal Institution for their support.

All correspondence should be addressed to Dr Graham Hollister-Short, Room 448, Sherfield Building, Imperial College, London SW7 2AZ.

The Contributors

Dr John Bradley,
4 Wellington Close,
Dibden Purlieu,
Southampton,
Hampshire SO45 4RL,
England

Dr Terry Gourvish,
Business History Unit,
London School of Economics,
Houghton Street,
London WC2A 2AE,
England

Professor A. Rupert Hall,
14 Ball Lane,
Tackley,
Oxfordshire OX5 3AG,
England

Rev Dr Richard L. Hills,
Stamford Cottage,
47 Old Road,
Mottram-in-Longdendale,
via Hyde,
Cheshire SK14 6LW,
England

Professor Ian Inkster,
International Studies,
Nottingham Trent University,
Clifton Lane,
Nottingham NG11 8NS,
England

Dr Alan L. Lougheed,
Department of Economics,
The University of Queensland,
Brisbane,
Qld 4072,
Australia

Dr Steve Travis,
51 St Lawrence Quays,
Salford Quays,
Manchester M5 2XT
England

Michael Wright,
Science Museum,
South Kensington,
London SW7 2DD,
England

Notes for Contributors

Contributions are welcome and should be sent to the editor. They are considered on the understanding that they are previously unpublished in English and are not on offer to another journal. Papers in French and German will be considered for publication, but an English summary will be required. The editor will also consider publishing English translations of papers already published in languages other than English. Three copies should be submitted, typed in double spacing (including quotations and notes) with a margin, on A4 or American Quarto paper. Include an abstract of 150–200 words and two or three sentences for 'Notes on Contributors'.

It would be appreciated if normal printers' instructions could be used. For example, words to be set in italics should be underlined and *not* put in italics. Authors who have passages originally in Cyrillic or oriental scripts should indicate the system of transliteration they have used. Quotations when long should be inset without quotation marks; when short, in single quotation marks. Spelling should follow the *Oxford English Dictionary*, and arrangement H. Hart, *Rules for Compositors* (Oxford, many editions). Be clear and consistent.

All papers should be rigorously documented, with references to primary and secondary sources typed separately from the text in double spacing and *numbered consecutively*. Cite as follows for books:

> 1. David Gooding, *Experiment and the Making of Meaning: Human Agency in Scientific Observation and Experiment* (Dordrecht, 1990), 54–5.

Subsequent references may be written:

> 3. Gooding, *op. cit.* (1), 43.

Only name the publisher for good reason. For theses, cite University Microfilm order number or at least Dissertations Abstract number. Standard works like DNB, DBB must be thus cited.

And as follows for articles:

> 13. Andrew Nahum, 'The Rotary Aero Engine', *Hist. Tech.*, 1986, 11: 125–66, p. 139.

Line drawings should be drawn boldly in black ink on stout white paper, feint-ruled paper or tracing paper. Photographs should be glossy prints of good contrast and well matched for tonal range. The place of an illustration should be indicated in the margin of the text where it should also be keyed in. Each illustration must be numbered and have a caption. Xerox copies may be sent when the article is first submitted for consideration.

Theory and Responsibility in Science and Technology

A. RUPERT HALL

Is man the only proper study for mankind? Alexander Pope's celebrated dictum was famously endorsed by Dr Johnson, who declared that 'the knowledge of external nature, and the sciences which that knowledge requires or includes, are not the great or the frequent business of the human mind'. More essential to human wisdom and welfare were knowledge of right and wrong, of religion and of history: 'we are perpetually moralists, but we are geometricians only by chance'. The study of astronomy or chemistry could be at best a harmless avocation.[1]

Until the early years of this century at least, Johnson's opinion was shared by many influential figures. Insofar as it shaped educational policy in the UK it was eroded not so much by argument as by brute practical necessity: technology had become a great force in the world and (in some way not wholly transparent) technology sprang from natural science. I doubt whether the changes that have occurred in the university and school curriculum of the UK since about 1850 commended themselves on purely intellectual or moral grounds to the majority of those who, in the end, brought about their introduction.

For a century and a half we have groped our way to what is intended ideally to be a coherent, progressive system of education at primary, secondary and tertiary levels, with science and technology now introduced to tiny tots. It is hardly surprising if, in the course of this evolution, many have come to believe that the *only* justification for teaching and research in both science and technology is utilitarian: a point of view that Johnson would hardly have comprehended, any more than he could understand how a bicycle might be superior to Shanks's pony as a means of locomotion. The widely repeated argument runs that we must educate engineers and technologists for our commerce, industry and national defence; and that behind these practical men and women, to sharpen their tools as it were, society needs scientists too. The idea of an education to enrich the life of the individual has no place in this argument, for reasons so obvious that I need not examine them.

I hasten to add that Johnson, like so many of his time and before, was quite aware that the study of nature could bring intellectual satisfaction, a point wholly ignored by the utilitarian argument. Our attempts to understand the universe have been defended, and well defended, on moral, theological, intellectual and indeed aesthetic grounds for many hundred years, if not millennia; though not infrequently in conjunction with utilitarian appeals to the usefulness of astrology or the botanical pharmacopoeia. Arguments similar to those used to defend the study of nature could also be applied to the improvement of its exploitation. But I need hardly recall now the names of such brilliant expositors of natural science and of the delights of studying it, within our own century, as Arthur Eddington and Peter Medawar.

The intellectual delights concealed in science may be less accessible than those of literature, or those perceived by the eye and the ear. And they may generate a somewhat snobbish attitude: 'I pursue science for the sake of mental bliss, you as a technologist merely exploit nature for gain.' There is a story told of the American historian of science of the last generation, George Sarton, who when once challenged as to why he did not pay more attention to the development of technology, replied: 'I am interested in medieval thinkers, not medieval tinkers.'

There may be those who judge it necessary to draw a distinction of kind between the discovery of new understanding of the universe and the discovery of some useful new property of its fabric; say, on the one hand, Chadwick's discovery (1932) of the neutron and the discovery of the transistor effect (1948) by Shockley, Bardeen and Brattain. Both discoveries resulted from purposeful research programmes: one gave real existence to a predicted entity, the other made neglected properties of known entities useful. The obvious distinction between concerning oneself with *understanding* and concerning onself with *properties* is that new understanding entails the confirmation, extension or modification of theoretical structures – today, increasingly, mathematical – while a new property may simply have lain, as it were, hidden within that structure. Marconi's insistence on investigating the propagation rather than the physical character of spark-generated electromagnetic waves is a case in point. Of course, it is also obvious that the revelation of some fundamental new property in our world by experimental technology may enforce modification of our understanding of it; Marconi's work again affords examples. For this reason, and also because of the vast increase of sophistication and theory in contemporary technology, the sense that understanding is more noble than exploitation has been much weakened. Moreover, few can now glorify the former in terms of a God-like perception of eternal verities.

I have now begun to consider the relative interest and relative value of science and technology, of understanding nature and mastering nature. It is beyond my present purpose to say more about the 'absolute' value of either (if it makes sense to invoke such a thing); but perhaps before proceeding I should explain further the difference (as I see it) between learning to understand nature and discovering its properties. Understanding, which can only be in terms of the philosophy, language and intellectual context of the time and so may require to be modified or subjected to restrictions at a later time, is universal: it says how things are (or, if you prefer, how they are seen to be). If we write Newton's law of gravitation in the form

$$f = \frac{m_1 \ m_2}{d^2}$$

we express in symbols a fundamental understanding of force, but the expression does not of itself enable us to calculate (in some suitable units) the force tending to unite two known spherical masses whose centres are separated by a known distance. On the other hand, statements about the properties of nature invoke particular classes of things and their characteristics. Thus, Maxwell's propositions about the existence of electromagnetic radiation in space added to understanding, while Hertz discovered the property of this radiation, as produced by an electric spark, that it behaves experimentally like light – it can be reflected, refracted or polarized. Note that this property was not at all for Hertz and for those who soon joined him in exploring the new phenomena a property that was obviously useful. This was to be the distinct apprehension of Marconi. Marconi initially discovered nothing new in physics; he correctly appreciated the communication potentialities of wave propagation.

I hope it may be agreed that the correct response to the attainment of new understanding of nature is: 'How fascinating! how greatly this enriches/completes/modifies our theoretical picture!' To say 'How useful!' or 'How can we make this new understanding useful?' is to invite Faraday's well known response to that inappropriate question. When a new *property* of things is revealed, as with Hertz and electromagnetic waves, usefulness is not necessarily apparent and is certainly not necessarily significant. On the other hand, it may be (as with the transistor) that the property has been sought precisely because its usefulness (if it could be identified) had been foreseen, in which case the response 'How useful!' is certainly appropriate.

To return to the issue to which I was approaching before – that is to say, the distinction between science and technology – I suspect that for many people today there is none: they perceive a continuous spectrum of knowledge and competence, stretching from mastery of applied mathematics at one end to changing the oil in your car at the other. Even the British Association seems to find no need to separate pure science, applied science and technology – this last a vast range in itself. This comes about not only because of the intellectual interpenetration of science (dealing with ideas) and technology (dealing with things), but because science, like technology, has become much concerned with management. The institutions of 'big science', and indeed any university department, of course have to be managed. But in addition, very many scientists are essentially concerned with management – the management of the environment (which we call conservation), of industry or of health, for example. And management necessarily has a purpose: it pursues an object of some kind, not the discovery of new truths as such. Science of a rather humble kind tells us that the natterjack toad is not the same animal as the common frog; its students may also tell us that either or both of these species occur less commonly than they used to do. But the purpose or objective of increasing or diminishing the number of amphibians is a *human* purpose, outside science, outside considerations of truth or falsity, and is a managerial purpose. To make an inventory of our planet (from which we may learn that such birds as the great auk and the dodo are now extinct) is scientific; to have feelings about the disappearance or modification of species, or their management for objectives that humans determine, is managerial.

So, as science has come to be more involved in human considerations, to be directed by considerations that are essentially moral or aesthetic, and therefore more concerned with human affairs, it has inevitably become entwined with technology, which might be defined in a very general way as the management of the properties of nature for human purposes. There is no way in which we can manage nature for nature's purposes, though some may mistakenly seek to do so.

Not so long ago the distinction between 'science' and 'technology' was quite clear to most people; science (I always mean natural science) was concerned with knowledge of the natural world, while technology was knowledge of the various crafts and processes by which men exploit the natural world in order to satisfy their needs. Indeed, it is not much more than a century since the word 'technology' ceased to bear an even narrower sense; the word originally signified writings about the crafts just as the word historiography signifies writings about history. The need for a word defining that general activity which is concerned with the design, preparation, manufacture and construction of all kinds of artefacts and facilities from pins to bridges, or, to put it another way, of all who are involved in the exploitation of materials both organic and inorganic, is of recent origin. In the past, people were content to use only specific terms such as coach-builder, weaver, farrier and so on. The production of food, goods and services was not regarded as a function possessed of a general character.

Further to emphasize the singularity of crafts, in contrast with the generality of technology, note that well into living memory many trades were concerned with effecting one step in a chain of processes, the shoe or plate or fabricated piece passing from bench to bench until it was finished. In all this machinery played its part, from the simple treadle-operated 'oliver' of the chain-smiths to water-driven saw, lathes and polishing-machines. The first great change from this type of production came in the late nineteenth century with the introduction of machines that could produce goods automatically from feeder stock, the second change when such machines (and other far more complex ones) were controlled by punched tape or computer. Obviously, the tendency for the future is for more and more goods and services to be provided by machines – mostly involving electronic devices – while the human element in technology is limited to the design and in part the construction of the machines that make things. Only very large, singular artefacts like bridges, dams and motorways are likely to adhere to a more traditional pattern; in general, technical activities such as these which are labour intensive change more slowly.

I pointed out previously that today's scientist is quite likely to be concerned with management – management of production, of the environment, of energy supplies, of health systems. Similarly, the modern technologist is concerned with the management of both machines and people, since few enterprises can completely dispense with a labour force. Recall that when the spinning and weaving of textiles was first automated more than two centuries ago, it was found necessary still to employ very many hands to repair breakages, to monitor the quality of the product and to feed the machines with raw materials. The need for this type of intervention remains but has been much reduced by the introduction of more complex machines. But my chief point is that again science and technology seem to converge in comparable managerial functions, moving away from their root functions of creating new understanding of nature and exploiting the properties of nature.

Now, of course, we know that the alliance or symbiosis of science and technology is an old one, and that it is deeply characteristic of modern European civilization. Cultures other than our own have found that it was proper and useful to put their theoretical and especially their mathematical exploration of the world to the service of religion, and perhaps as an extension of that to use it in agriculture and other practical activities, but no other has so closely connected its theoretical investigations with its arts and crafts. The historian can trace the evolution of this process during the Middle Ages and early modern times, the development of a fruitful relationship between intellectual and manual activities, and the recognition of this relationship as a desirable objective, to be encouraged and furthered, during the seventeenth century particularly. History shows, for example, that both the Royal Society of London and its opposite number in Paris thought it fitting for philosophers to become cognizant of craft methods (and to record and analyse them), and that such scientific societies believed that philosophical – or in our language scientific – innovations could be of benefit to the practical arts, navigation not least, and also to the profession of medicine. It is well known, for example, that there was great contemporary interest in the derivation of new medicaments from the newly found plants of the Americas and the Far East; again, it is clear that the successful eighteenth-century steam engine is linearly connected with the experimental air pump of the previous century.

To recognize such relationships is not to allege either that science was in the West the creature of technology, or conversely that science alone generated an increasingly complex technology. Both these extreme positions have been maintained; the truth seems more complex. It is enough for the present to emphasize that through the centuries, and especially during the past two or three hundred years, the relations between science and technology have altered and that they have never followed any consistent pattern. Further, it is obvious that only in the nineteenth century did the more sophisticated and wide ranging types of theory, such as thermodynamics, begin to find technological and engineering applications. The theory of structures offers another illuminating example here, and one with rather remoter origins.

It is important to observe that because science and technology – different facets of the interface between humanity and the natural world – have many connections and relationships, because each has given opportunity and inspiration for achievement in the other, it does not follow that they are of the same character or that both address themselves to the same problems. Rather, they are complementary, as sense and metre are complementary in a poem. Verse which is all meaning or all music can hardly attain to the highest poetic art.

In my view there is a fundamental difference between *knowing* and *doing*. Obviously, the intellectual and the executive functions do combine, nevertheless. Each one of us *knows* things – indeed, we are all unique possessors of some information – and each of us *does* things; we may even say that we can do certain things only because we *know how to* do them. We may remark of the potter or boat-builder in a primitive society that he knows his timber or his clay well. So there can be no dispute that even the most elementary manual technology embraces knowledge, though it may be (as it is with the bower bird or the bee) knowledge that cannot be communicated by writing or speech. Equally, the most simple and early knowledge of nature known to us embraces an element of *doing*, if only the activity of observing and making records (for how can there be any knowledge without records of some kind?). But no one would argue that the Babylonian astronomer and the potter who prepared the clay tablet upon which he impressed his symbols for record were engaged in the same or even similar activities. Nor can 'knowing how to ... ' imply command of abstract, rational or mathematical theories concerning the technique in question. I would add that it is intuitively obvious that the work of the head differs from the labour of the hands, however closely these diverse organs may co-operate towards a common goal; and that the human brain has developed through the centuries (more than once, in the parallel evolution of cultures) general definitions, concepts and theories relating to the knowledge of nature – theories, explanations, pictures, call them what you will – concerning both external nature (the context of our being) and our own, internal, human nature.

Obviously, not all knowledge is of nature or of man: some of it is devoted to the gods, some to such abstractions as number, some of it is philosophy and metaphysics. And that such knowledge not of nature may have some possible relevance to technology I need not deny. Here I claim only that the pursuit of knowledge of all kinds, including the knowledge of nature, has in all societies been distinguished from the practical mastery of nature; that is, technology. The reason for the distinction is obvious: a man can very well exist, can make and can do, without this abstract kind of theoretical knowledge whose natural expression is in language, be it in words or in numbers. It may indeed be that some religious ceremony or other must precede the firing of a kiln or the launching of some new built ship, but pots can certainly be made without skill in chemistry and boats built and sailed by men ignorant of the sciences of materials and mechanics, or of geography and navigation. On the contrary, although through the long centuries of their early development mathematics and astronomy were closely bound to religion, their first stages had no relation to practical affairs, except for the business of land measurement.

To look at the matter from the other side, when we move on from prehistory to the oldest technological writings, those of classical antiquity, we find emphasis on practice rather than on theoretical understanding. For the most part the ancient authors were content to explain how things should be done in order to achieve stated results, avoiding (or at any rate not recording) discussion of general theories by which the procedures described might be justified and accounted for. Even what we might call general engineering principles are usually left unstated. We have to infer that these ancient authors took little interest in (to use our terms) the strength of materials or the theory of structures, hydraulics or ballistics, save in making specific recommendations. Yet they were much concerned with the design of structures and machines in terms of aesthetics, endurance and utility. One of the earliest and best known, the pseudo-Aristotelian *Mechanical Problems*, successfully interprets more complex devices such as the pulley and windlass in terms of the lever, but only feebly accounts for the action of the lever itself. Centuries later, Hero's *Pneumatica*, a practical description of automata and other mechanisms, opens with a preface on the nature of the vacuum, but reference to this in the text occurs once only, in the treatment of the siphon.

Leaping forward again to the first printed literature of technology in early modern times, one finds that however fanciful some plans for machines or cities may be, the essence of the description is practical; authors did not trouble their readers with the philosophy of the machine or of town planning. Engravings were made pleasantly symmetrical by geometry, but there was little or no quantitative computation. The higher, mathematized theory of technology makes a late appearance and is science-dependent.

With apologies for this superficial historical sketch I come now to a moral point for which it prepares the way. If we don't maintain a distinction between science and technology we lose that between knowing and doing. That is to say, we may be slipping into the view that our knowledge empowers us to do something or other, that because we *can* do it we *may* do it, and that (finally) since we may do it we must do it. Now in the ordinary terms of living we none of us accept any of this series of propositions as valid. In ordinary life we do not exploit all the knowledge that we have, and we certainly do not do all that we know how to do. We classify as psychopathic those individuals who cannot distinguish between *can* and *may*, and who are unable to refrain from taking advantage of knowledge or power that they possess.

If this self-denying principle applies in the ordinary affairs of life among decent people, does some similar principle apply to the world of engineering and technology? It is clear that it *ought* to apply, and that it would be prudent that it *should* be applied. There are obviously three chief types of case to consider: take first those possible instances in which a technological innovation seems to work against the whole of mankind, perhaps even the whole biosphere. Let me offer one example from antiquity. Archaeological evidence suggests that the new knowledge of the 'parting' of silver from lead ores, and its exploitation by the Athenians at Laurion in particular, may have released enough lead into the atmosphere during classical times to poison at least the whole of Europe to a mild extent. I need hardly recall more specifically than by citing the words 'acid rain' that a warm debate continues about the perilous significance and arguable morality of recent and even vaster industrial activities in modern times.

Second, there are cases where innovation works to the advantage of one section of mankind, but to the disadvantage of another section. For example, the successful introduction of synthetic rubber depressed the market for natural rubber, of which the growers in various primary producing countries had formerly enjoyed a monopoly. Third, there are cases where the damage is suffered by individuals or small groups, such as workers in a dangerous trade or those afflicted by avoidable accidents, such as the explosion of a steam boiler. Clearly, damage on each of these three scales of magnitude could be totally avoided if the innovations were rejected; damage may be obviated, or at least minimized, by prudential and restrictive measures. Prohibitions on industrial practices grossly endangering health, or perpetrating fraud, have been imposed in civilized states since the early part of the nineteenth century; the regulation of potentially dangerous technological innovations has been increasingly a feature of the twentieth century. Yet the determination of what is good and what is harmful in industrial methods is not always easy, since (for example) what is good for the individual may be, at least in the short term, damaging to the community. The near or total disappearance of such arduous and dangerous trades as iron-puddling or coal-mining has invariably entailed social disruption.

A 'counter-factualist' historian might examine the effects on world history of a deliberate choice – never in fact on the cards – to eschew the successive uses of fossil and nuclear energy: concentration upon the technology of renewable energy sources could, perhaps, at a slower rate, have brought world energy use to something like its present magnitude; one can perhaps even imagine jets powered by alcohol or some other liquid fuel of biological origin. But the historical fact is that the history of the vast expansion of energy use in the past five centuries – if we set aside nuclear power – shows a long series of small-scale, mostly private, innovation by multiple stages, each seeming to be in the direction of financial economy, productivity and convenience. Coal was cheaper than wood, oil more convenient (and often cheaper) than coal. There was a quasi-democratic natural selection constantly favoured by technical evolution. Accordingly, it is often suggested that no technological innovation promising profits has ever been neglected or set aside – that men have always taken the next step offering to take them nearer to the goals of wealth and power. Unsuccessful innovations, like the hot-air engine, though practicable and even finding some application, have ultimately failed because of their limited range of advantage or because they did not suit the financial and industrial structures of the time. It isn't easy to think of innovations that have been *suppressed* because of their damaging effects upon the whole or some large section of mankind, partly because innovations are so often well established before their damaging consequences become clear. Some inventions, as Bernard Shaw tells us, may have been suppressed by particular interests for their own protection, but that is a different matter. Again, Nobel believed dynamite to be so terrible an explosive that it could not be used in warfare, but he manufactured it for industrial use nevertheless. And the sceptic will claim that no weapon of war has seemed so terrible that it is not used, if circumstances seem unambiguously to favour the user of it.

In civil affairs, our inability to conceive of clear limitations upon technological change arises, I think, not so much because of ignorance or our consciousness of moral weakness as from lack of proper authority. We can hardly, as yet, imagine any mechanism by which a decision to retreat on a global basis from the use of nuclear energy for power generation might be framed, still less enforced. Even in the form of a decision taken by a single nation such abnegation seems too quixotic. The world is capable of making general but ineffectual resolutions to control rather small-scale activities like whaling, but can one yet imagine a global limitation on the use of the internal combustion engine, or even global control of the pollution it produces? It is one thing to move, if we can, towards cheaper, more efficient and more environmentally friendly engines, but quite another to halve the power output of the engines we have. Moreover, the criteria by which this or that technological step may be judged beneficial or harmful to people are, as I have said, by no means always clear to everyone. For example, consider the kind of agrarian self-interest which in the past has produced both swamps and deserts, and our present need to make a choice between nuclear power stations and windmills. 'Green' is not necessarily beautiful to a rambler.

That, on a global basis, there could be any review or process of certification of major technological changes before their large-scale exploitation seems hopelessly idealistic. We have always become aware of the good or bad effects of innovation when it is far too late to choose between fertilizing the young shoot or nipping it off. On the international scale we can only look forward to a continuation of the present haphazard and ineffectual state of affairs, though on the national scale we may expect the technical policies of governments and corporations – whether these be conventional or innovatory – to come increasingly under critical examination. However, when we turn to the lowest level of consideration it is clear that great progress has been achieved in historical times, and continues still. All civilized countries have in force a mass of legislation concerned with the protection of workers in industry and agriculture, and the protection of the public at large in their role (voluntary and involuntary) as consumers. Much of the legislation concerned with the environment (for example, with clean air, clean water and clean beaches) is in fact aimed at the protection of the human population, rather than directed to the benefit of animals and plants. This shielding of mankind from the ill effects of industrialization, and of the population pressures created by industrialization, has been brought about by two centuries of social pressure and a great deal of fundamental scientific and medical investigation. Because of such research – think how little was known of the vectors of transmitted disease only a century ago – and because of the zeal of reformers, society has been compelled to assume responsibility for the effects of industrialization and congestion in advanced countries. Free action in all respects is no longer possible. Free choice in all matters cannot be tolerated. Responsibility consists of the voluntary surrender of the freedom to injure others, even unwittingly.

Let me make two points about this long process, one encouraging for the future, the other less so. Take the former first: a great deal of the exercise of responsibility by engineers and technologists has been self-assumed; that is, the professions have regulated themselves, have defined the forms of responsible action and enforced them. In Iceland, electrical engineers resolved to protect their native tongue by eschewing technical terms of foreign origin, and to employ only neologisms coined from Icelandic roots. In England we may regard the growth of engineering institutions as representing, in part, the exercise of corporate responsibility; among their objects was the protection of the public from the disasters which might occur from the use of untraditional means of transport and of construction. The public has by no means only been protected by the power of the state, which is neither always sensitive nor always prompt. Second, the relation between the rate of technical change and the growth of responsibility among those bringing it about is not simple; notably, the social response is rarely triggered by a single technical innovation but is a response to a complex of many. I am not referring of course to simple cheats, like putting ground chalk in bread flour or other such examples of food adulteration, though these have been serious enough issues in their time. I mean that if you consider a simple concept, like 'a steamship', it proves that responsibility, protection for passengers and crew, has to extend in a host of directions, from the qualification of the officers to load lines, from boilers to lifeboats, from the accommodation of passengers to navigation equipment, all of which was in the days of sail simply left to the discretion of ship owners. Here and in a myriad other instances technical change in time enforced a new development of social responsibility.

We now expect in the civilized world engineering, industry, manufacture and technology in general to be activities that are carried on in a responsible way, and in fact we hedge them and their practitioners about with thickets of legislation and codes of practice in order to ensure that this is so. *Doing* has to be responsible, even if what you are doing is only walking your dog or getting rid of garden rubbish. Now I ask: what about *knowing*? The formal tradition of the occident – often abrogated in fact, one must admit – has been that knowledge should be available to all, and that the search for new knowledge should be open, free and unrestrained. One perhaps over-simple view of the history of natural science portrays it as an ascent to freedom, to intellectual emancipation, to the opportunity to frame and teach strange and complex ideas that seem genuinely to correspond to the subtle and complex nature of things. In the period from about 1880 to the outbreak of World War I – the age of Rutherford and Wells – in which public esteem and admiration for science were at their height, such a high view of its liberalizing and virtually infallible character was almost universal among people of a liberal outlook. That there was never, in any period, an unrestricted freedom of investigation into nature – for the few with the leisure and means to undertake it – is, however, obvious. While we note that the natural and religious repugnance felt for dissection of the human body was replaced (near the end of the nineteenth century) by an equally strong, and in some countries highly effective, feeling that the use of animals in vivisection or indeed any experimentation was intolerable, we must also record a growing recognition of the dangers linked with some avenues of research. The study of particulate radiation offers a telling example, while (in moral terms) what one might call the 'straight scientific' version of Darwinism never lacked critics who spoke with weight and authority.

My two examples illustrate the need for a new distinction. At this point in the discussion we need to separate the simple right to think and to propound from the more complex right to investigate experimentally, perhaps in ways that may be damaging to others, whether people or animals or the environment. It is morally right and absolutely necessary that the right to experiment should be properly regulated – again, what is at stake is a form of action and action (all must agree) should always be responsible and prudent rather than impetuous. It is neither fitting nor feasible to impose restraints upon the right to theorize scientifically but, human nature being what it is, one dare not declare that the right to experiment is equally inviolable. The right to think is inviolable; the right to know – to confirm theory by experiment – is of a very different character because it surpasses the sphere of the individual and may trespass on the rights of others.

This restriction being allowed for, and it being understood also that the knowledge to be sought is of a universal and impartial kind, is it reasonable to suppose that there exist general reasons for arguing that the search for such knowledge ought in principle to be subjected to restraint or limitation, as I have argued that technological applications must necessarily be controlled and restrained? We may, for instance, find that the answers to some questions that scientific research may propose are too costly to obtain, at least by the proposed experiments. This matter was dealt with by John Ziman in a recent issue of *Science and Public Affairs* (Spring 1994, pp. 45–7). Again, a project might be of so destructive a nature that no volunteer should be permitted to pursue it, or in some other way might be judged extremely hazardous. But whereas expenses can be computed with some exactness, dangers (of whatever sort) are necessarily debatable, and commonly (once more) arise not so much from the gathering of new knowledge as from its application. For example, ought research on particle physics or organochloride compounds such as DDT to have been discouraged in the 1930s because of possible future perils, even if these could have been at all precisely foreseen? For myself I cannot believe that such a policy of timidity makes sense, for since it is in principle impossible to foretell the consequences of the unknown, to adopt it would in logical extension be to declare that we must abandon all fundamental scientific research.

This point in effect returns us to an earlier point: it is the application or exploitation of knowledge in physics or chemistry that may have fearful consequences (or be suspected of bringing such consequences), not the knowledge in itself. To quote a writer to *The Times*: 'Science advances knowledge. It is engineering which converts the knowledge into useful artefacts' (E. Duchworth, 4 August 1994). Equally, it is (or may be) engineering that causes human or environmental disasters, or at the least creates the means for them. Therefore it is essential that the functions of application, through engineering, be handled responsibly. Theory as such can be neither responsible nor irresponsible since it is an intellectual structure only, changing nothing in the external world.

It is upon this issue, the issue of responsibility, that the necessity to understand and maintain the distinction between science on the one hand and technology on the other turns. If we do not, we place ourselves in the position of journalists (and, it must be said, of many others who ought to know better) who see 'science' as responsible for everything they do not like in our modern world, or conversely hold technology to be inevitable and irresistible in the changes it effects. Such people may then say: 'Scientists should be stopped from doing these frightful things', thereby (in wishing for greater responsibility in engineering and technology) imposing a restriction on the right to know. What is at stake has been a fundamental attribute of our evolving civilization, the freedom to theorize freely on the one hand, combined with responsibility in creating and acting on the other.

Notes

An earlier draft of this chapter was read as the Presidential Address to Section Q of the British Association, during its meeting at Loughborough in September 1994.

1. The quotations are from Johnson's 'Life of Milton' in *Lives of the Poets.*

On the Lift Pump

M. T. WRIGHT

INTRODUCTION

The lift pump, or as some writers have called it, the suction pump or the suction lift pump, appears in the historical record quite suddenly, and fully formed, in the early fifteenth century.[1] There had existed since Hellenistic times another type of pump for raising water, the force pump. The lift pump and the force pump remain the two principal types of reciprocating pump. The question has been asked as to how the lift pump came into being, whether as a development of the force pump or as an independent invention.[2] I offer here a small contribution to this debate.

At first sight the two pumps are similar. In both there is a cylinder, the *barrel*, in which another member slides; the latter is often called a piston, but in the lift pump it is properly called the *bucket*.[3] In the force pump it may indeed be a *piston* (a spool-shaped piece fitting the barrel, fixed to a narrower rod by which it is worked) or it may be a *plunger* (a component of uniform diameter throughout its length), depending on the design. Other names for these components, such as sucker (for bucket) and forcer or ram (for piston or plunger) will not be used here. Both pumps usually have at least two valves, sometimes called clacks or, especially in some old texts, and rather confusingly, suckers; I shall call them *valves*. Despite these similarities the two pumps work in rather different ways, which will be discussed below.

MECHANICAL PRINCIPLES

It seems necessary to make some preliminary remarks concerning water and air because they have an important bearing on how water pumps work, and because some writers on the history of pumps seem to have been unclear about them, at least by implication. These points may not have been, indeed almost certainly were not, fully understood by those who invented the pumps that we are considering, but they are important to us in forming a understanding of how each type of pump works and an appreciation of its limitations; and these things should be grasped before one ventures into speculation about the development of pumps.

It is widely understood that water is so nearly incompressible as makes no difference for our purposes. Similarly, if an attempt is made to rarefy water, the liquid will not increase noticeably in bulk but the air, which water usually contains in solution, will tend to come out of solution and collect at the top. The air space will also contain water vapour up to a pressure corresponding to the ambient temperature. At ordinary temperatures this pressure is small, but it rises to atmospheric pressure as the temperature approaches the normal boiling point. If one lowers the pressure over the water to the vapour pressure for that temperature, the water boils at that temperature; that is, bubbles of vapour form within the bulk of the liquid. The formation of such bubbles or cavities in the water is known as cavitation. The air and water vapour are elastic and can be dilated appreciably.[4]

Next, the pressure at any depth under the surface of still water is the sum of the hydrostatic pressure (proportional to the depth) and whatever pressure exists at the upper surface of the water. This pressure is exerted uniformly and normally[5] on the containing walls at that depth, whatever the shape of the container. In practice there may in addition be a pressure due to the flow of the water, but for our purposes (in pumps in which the water moves quite slowly) it may be ignored. The maker of pumps, even simple ones, does, however, have to contend with shock, derived from the momentum of an appreciable mass of water suddenly arrested even from no more than a moderate velocity.[6]

These principles are sufficient to equip the reader to spot fallacies and omissions in some descriptions that may be found in print of the way in which water pumps work; I now offer my own explanations of the two pumps mentioned in my introduction.

ACTION OF PUMPS

I describe first the working of the earlier machine, the force pump (Figure 1).[7] Inlet and outlet, both equipped with non-return valves called respectively the foot valve and the delivery valve, both communicate with one end of the barrel, the other end of which is closed by the plunger. The plunger is partially withdrawn, so that water enters the barrel through the foot valve. The plunger is then pushed back, expelling the water through the delivery valve and into the pipe beyond.

Note that the water may be said to be drawn into the barrel by suction.[8] There is no reason in principle why such a pump should not be set above the water level and made to draw water up a suction limb.[9] But the foregoing elementary explanation of how the pump works assumes that the plunger and valves do not leak and that no air accumulates in the pump. For a real pump some further points should be made.

The performance of any reciprocating pump is always diminished by water flowing back through the valves as they close, together with leakage through the valves when 'shut'. In the force pump, leakage of water past the plunger during the delivery is another source of loss.[10]

The behaviour of air in the pump also affects its performance and may play a more drastic role in limiting its application. The design of the pump may be such that some air is trapped inside it to begin with. Moreover, as water is sucked into the pump, air in the water will tend to

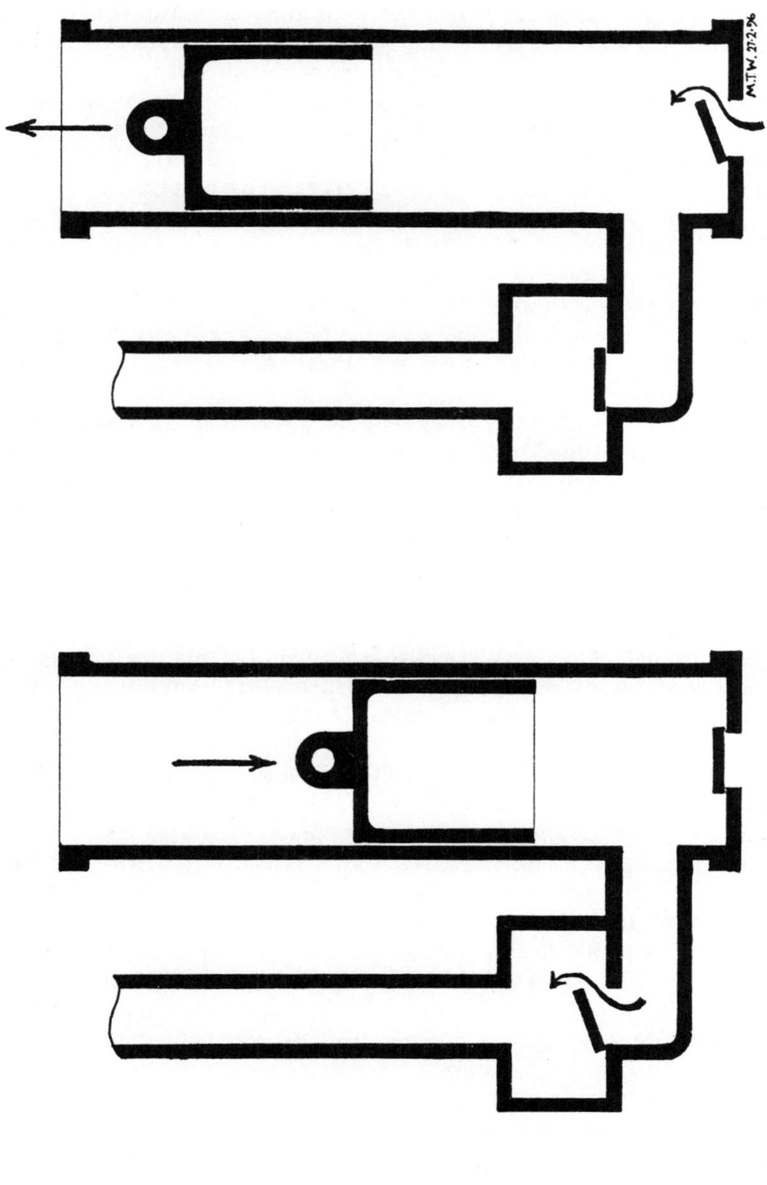

Figure 1 Force pump.

come out of solution. Some air may go back into solution when the pressure is raised on the delivery stroke, but usually it does not stay in contact with the water long enough for all , or even much, of it to dissolve. The presence of air diminishes both the suction achieved and the volume of water drawn in because it dilates as the plunger is withdrawn.

The arrangement of the pump may be such that the air is expelled with the water, but otherwise there is the possibility that air will accumulate in the pump, progressively diminishing the suction as the plunger is raised. With a particularly unfortunate combination of arrangement and proportions, this could hamper the working of the pump altogether.[11]

Moreover, if the pump is set with the top of the barrel above the water level, air may enter by leakage past the plunger on the suction stroke and so reduce the suction achieved. This is a potential source of really serious trouble.[12] If the plunger were very leaky, water would enter the barrel only to the height of the water surface outside.

A pump with these faults may stand a better chance of giving satisfactory service if worked drowned, set below water level. I shall return again to this point.

A final point is that the pump may be arranged with either a plunger or a piston, working in any direction, up, down, horizontally or obliquely; as to general principles, it makes no difference. A particular arrangement may, however, be chosen in any application for practical reasons such as the ease of venting air, or to control leakage, or for convenience in working it.

We now consider the working of the lift pump (Figure 2). This has a foot valve admitting water to the barrel just as in the force pump, but there is no branch with a delivery valve leading to a rising main. Instead, there is a valve or set of valves in the reciprocating element itself, the bucket, and the barrel is extended upward to the level at which the water is to be discharged.

Water is drawn into the barrel by suction, just as in the force pump, as the bucket is raised. Simultaneously, water above the bucket is lifted up the barrel. When enough strokes have been made to fill the barrel to the outlet, water is delivered during this stroke. On its downward stroke, the bucket is idle; the valve in it opens so that water passes freely and the bucket merely sinks. (If it is less dense than water, for example if it is made of wood with but few metal fittings, it has to be pushed down.)

Note that the water does not descend with the bucket; still less does the bucket fall with a blow on the water below. Neglecting leakage back through the foot valve, on the down-stroke the water above the bucket should actually rise, and water should be discharged, because it is displaced by the bulk of the pump rod (or rope) attached to the bucket as it is plunged below the surface.[13]

It is obvious from this description that the lift pump can work only with the barrel rising from the foot valve at its lower end. Usually it is set upright, but it may be used obliquely and this is done where space is limited.[14]

In practice the lift pump suffers from the same losses due to leakage at

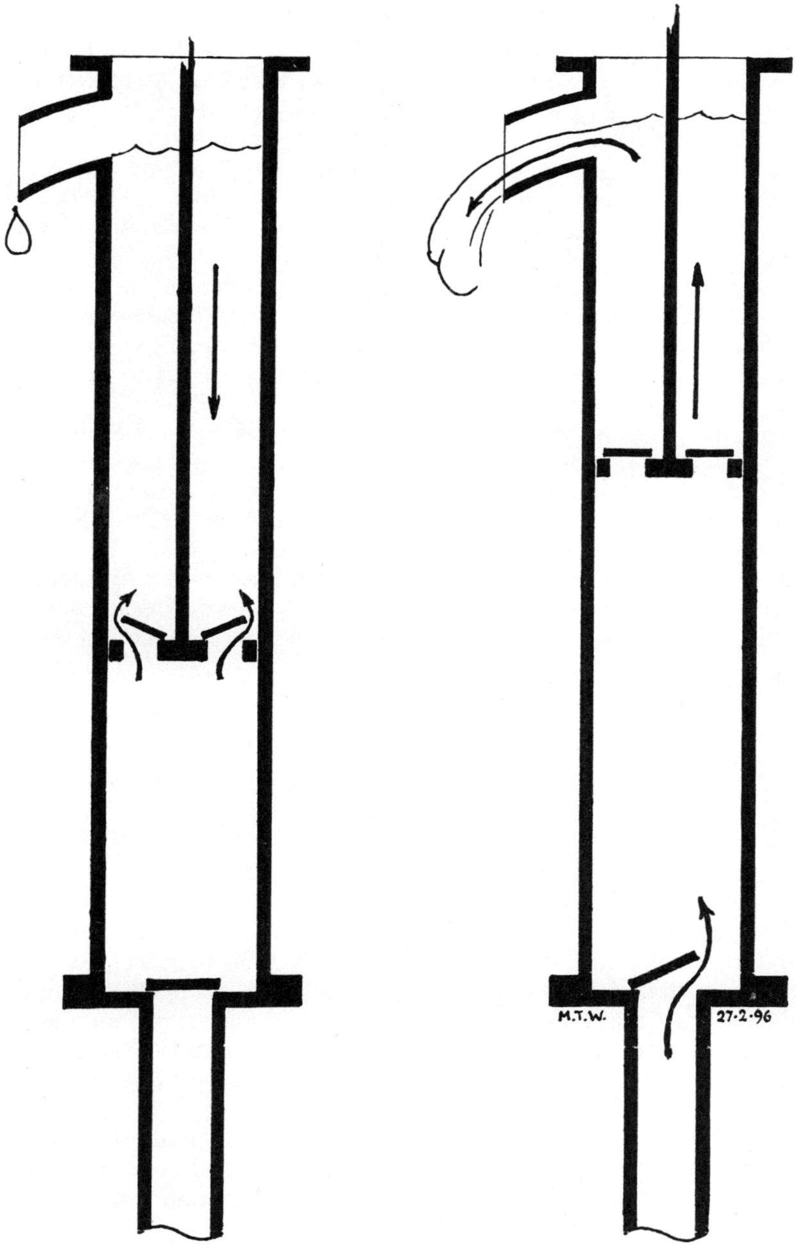

Figure 2 Lift pump.

the valves and their action in closing as does the force pump. The delivery is further diminished by leakage past the edge of the bucket. But if we consider the effect that this leakage can have on the suction in the space below the bucket, once there is a little water above the bucket such leakage is only the ingress of water, not air; so the suction of the lift pump is not liable to be spoiled by the leakage of air as that of the force pump may be.[15] It must be observed, however, that it can be difficult in practice to ensure that the pipes themselves and any joints in them are proof against the leaking in of air.[16] On the other hand, any air below the bucket, whether coming out of solution or getting in through leaks, rises to the bucket and passes it, and so is purged, at the beginning of each downward stroke of the bucket.

It is this combination of properties intrinsic to the design of the lift pump – that leakage past its reciprocating part does not allow air to spoil its suction, and that air cannot accumulate in it stroke by stroke – that makes it suitable to be set above the level of the water to be raised, drawing water through a suction limb.[17]

Irrespective of the type of pump, it is well known that the height to which water can be raised in a suction limb is limited, and the theoretical limit is the height of a column of water that exerts a hydrostatic pressure equal to the pressure of the atmosphere, about 9 metres. It is less well appreciated that as this limit is approached – that is, as the hydrostatic pressure on the suction side of the pump is reduced – the problems of leakage in of air and of cavitation within the pump during the suction stroke become more insistent. In practice the performance of a pump drops away before the limit is reached.

HISTORY IN OUTLINE

The lift pump is first found as a scheme on paper, during a period for which we have no really firm evidence for the pumping machinery actually being used at all. Therefore, in this section I make a distinction between the appearance of the idea and the confirmation of its successful execution.

The lift pump is first soberly attested to as a certain and accomplished practical reality in Agricola's *de Re Metallica* of 1556.[18] Agricola states in his introduction his intention to be as strictly factual as possible,[19] and the machinery and methods of which he gives details may be taken to reflect actual mining practice in Saxony in the years between Agricola's going to Joachimsthal (as town physician) in 1527 and his completion of the text at Chemnitz in 1550. The lift pump is shown in several configurations, some performing heavy duty, and though the designs illustrated and described are in many respects coarse, they are utterly down-to-earth and practicable.[20] After Agricola, the use of this type of pump can be traced continuously to modern times.[21]

As a design, the lift pump can be traced backwards a century before Agricola, through drawings by Leonardo da Vinci and other Italian engineers, to its appearance in the book *de Ingeneis* compiled by 'the Sienese Archimedes' Mariano di Jacopo detto Taccola (born 1382) prior to 1433, in which lift pumps are included among several pump designs.[22] No earlier description or depiction of the lift pump has been found.[23]

Turning now to the force pump, the earliest description that I know confirming its actual use in post-medieval Western Europe is Bate's account of the machinery at London Bridge waterworks as he saw it in 1633.[24] It is possible, though perhaps unlikely, that what Bate described was the first plant that had been set up there in about 1582,[25] but it is also possible that what he saw was a replacement of the original plant to the same design. By the seventeenth century the force pump was in common use; Sir Samuel Morland made the important improvement of fitting a packing at the open end of the working barrel, making a seal around a plain plunger.[26] From then on it too has an unbroken history of use to the present day.

As with the lift pump, the force pump may be traced back as a design through various authors to Taccola. But we can trace an earlier ancestry for the force pump that the suction pump lacks, reaching back to Hellenistic times. Its invention is ascribed to Ctesibius of Alexandria (third century BC), whose own writings are lost but are alluded to by both Philon of Byzantium (third to second centuries BC) and Vitruvius (first century AD), who both give descriptions of the device. Another clear description is given by Heron of Alexandria (first century AD). A number of actual examples, more or less complete, survive from Roman times, some (corresponding to the descriptions in the literature) made of bronze and others made of wood.[27] The force pump was evidently one of those devices that continued in use and was carried over into Islamic culture, as we find a water-powered version in al-Jazarī, who likens the pump itself to a thrower of naphtha (Greek fire).[28]

Certain features of these ancient force pumps seem to be fairly constant throughout both the written and the archaeological record: they were either small or at most moderate in size, by our standards; they were at least usually arranged in pairs with a common delivery pipe, and were probably worked alternately to yield a reasonably steady delivery whether connected to a common hand lever or not; with one possible exception they had no significant length of suction limb. The bronze pumps appear not to have had any packing between the plungers and their barrels, and it is doubtful whether the packing of the wooden pumps would have remained very effective for long. Allowing the possible exception noted above, the evidence indicates that pumps were set with at least their lower ends below the surface of the water to be pumped, and it may have been essential to their adequate performance that they were fully immersed, as discussed above.

The range of applications mentioned in the literature comprises fire-fighting, supplying fountains (perhaps by raising water to a cistern at high level) and washing high vaults. The archaeological contexts of discovery suggest use for domestic and perhaps industrial water supply, bilge pumping and perhaps fire-setting (in mining). There seems to be a suggestion in the literature that towards the end of the Roman Empire the pump was commoner in the East than in the West, although all the recorded survivals come from the West.[29]

The importance of these considerations in the present context lies in any implication that they may have as to how widely used or how widely known the force pump may have been in the ancient world, which might suggest to us, in the absence of evidence, how likely it may be that the use of such pumps continued in an unbroken tradition through the medieval period. If the number of copies of the *de Architectura* of Vitruvius is any guide, the pump should have remained reasonably well known, at any rate to readers of Latin, even if it was perhaps not widely used.[30] It is at least interesting that Taccola illustrates what I shall call the displacing plunger pump (see below) in the same paired format, and in the *Codex Atlanticus* we find a design for a pair of lift pumps by Leonardo da Vinci, the configuration of which is very close to that of the ancient force pump. But whether the arrangement was borrowed or was reinvented, or the congruence was pure coincidence, it tells us little about the origin of the lift pump.

The alternative possibilities remain that the lift pump was in some way derived from the force pump or that it was an altogether new invention in or before the early fifteenth century. In the absence of further historical evidence, the best hope of making progress seems to be to follow others who have examined the context of its earliest appearance. And so we turn to the designs for pumps drawn and described by Taccola.[31] I do not reproduce any of Taccola's drawings here, for copyright reasons, but refer the reader to the published facsimiles as detailed in the references.

TACCOLA'S PUMPS

In his *de Ingeneis*, completed in 1433, Taccola gives drawings of a number of types of water-raising machines, including, as has been said above, the first representations of the lift pump. These designs are repeated in his later book, *de Machinis* of 1449. Some other designs for reciprocating pumps will be discussed here. There remain a variety of swapes and similar bailing devices, a bellows pump to which allusion will be made later and the earliest representations of the chain pump and what may be the diaphragm pump.[32] None of these need be discussed further in this context except to point out that their presence and variety indicate a considerable preoccupation with the raising of water. Sometimes a context is indicated, ranging in scale from the getting of water for domestic use to the working of water mills where there is no natural fall of water, and even perpetual motion schemes in which a mill pumps the water that is to drive it.

As to the reciprocating pumps, there is room for some debate about how each of them may have worked, if at all. Even so, I follow Hollister-Short's suggestion that those not clearly representing either lift pumps or force pumps may indicate transitional forms, so that it may be fruitful to consider them carefully; but I must state that, as drawn by Taccola, there are difficulties with all the designs.

We have here a compound problem. First, there are the conventions of Taccola's drawing; even with some machines that we recognize there are elements omitted that we must take for granted in order to 'see' the complete device. For example, when we look at Taccola's sketch of the 'complete' lift pump,[33] we see no clear indication of a foot valve, no packing around the bucket (which as shown would have to be a slack fit in the bore of the barrel, because the stirrup from which it hangs is fastened outside it) and no delivery spout. Without the first of these, the machine could only work in very limited circumstances as a bailing device;[34] without the second, it could not work at all; and without the third any water pumped up would spill uselessly back down the outside of the barrel. If we accept that we must supply these elements 'as understood', then there is little to be made, for example, of the fact that Taccola shows this lift pump without the suction limb with which it was usually associated later.

Second, there are palpable defects in Taccola's draughtsmanship. I do not refer to errors of proportion or perspective, which are hardly to be wondered at, but such errors as the connection of links at impossible points. For instance, in another drawing, of two lift pumps worked by a two-throw crank, one bucket is connected (as it should be) to a throw but the other is attached to the shaft between the throws.[35] Taccola is not alone in this type of blunder; such mistakes can be found even in the tidier, and so more plausible, sketches of Leonardo da Vinci; but errors like this lessen our faith in the reliability of the drawings generally.

Third, even where Taccola provides a descriptive text, it is usually most unclear whether we are presented with a record of an engineering success or an untried scheme. Indeed, we can only be certain on this point when we come to those that are clearly *not* workable, such as the schemes for perpetual motion. In others, such as where Taccola shows one man pumping water to drive a corn mill, our credulity as to whether the scheme is workable in principle is merely stretched.

Fourth, in other cases, in which we might argue that the drawing seems more convincing, and might even have been based on an actual and successful machine, we do not know whether Taccola himself was its designer or whether he drew a machine built by others. In the latter case we might excuse unconvincing details as perhaps the result of imperfect observation, understanding or memory.

In short, we do not know how much weight to give any supposed deficiency or error in any drawing, and there is certainly room for doubt as to how well many of these machines, as conceived by Taccola, could have worked. If, in the hope of rediscovering Taccola's intentions, we devise workable reconstructions, we risk 'improving' the machines out of all recognition and so creating a false history.

It is easy for us to be dazzled by hindsight in this way. For the historian of technology attempting to chart the course of invention, however 'inevitable' a machine or process may now seem, or however impossible a fanciful scheme, there are considerable difficulties in reconstructing the trains of thought that led to it. Nor is a logical train of thought necessarily a safe guide. It is evident that many inventions have been brought forward in spite of ignorance about how they work; even recent history is littered with instances in which real advances have been based on misunderstanding or faulty logic. And there is value to be got from reconstructing the mental processes that led even to a failure.

However, understanding the risk involved, I begin, as a mechanic rather than an historian, by applying to Taccola's drawings of pumps the hindsight of the principles that I have set out above. In doing this I am less concerned with how Taccola may have supposed that the pumps worked than with the actual behaviour of water and air in pumps made according to the drawings. Whether any problems identified in this process may represent actual defects in Taccola's designs, deficiencies or errors in his drawings and description, or errors in his understanding of workable devices that he observed or recorded faultily, I am less able to judge with any confidence; and so I leave the construction of hypotheses to others. But I am certain that this mechanical appraisal is a necessary preliminary to the erection of any historical edifice on the evidence of these drawings as a foundation.

I begin with the device that Hollister-Short names the 'solid ram pump with cannelure'; as explained above, I prefer to avoid the term 'ram', and for reasons that will appear I prefer to avoid the expression 'with cannelure', so I offer a new name: *pump with displacing plunger.*[36] Taccola shows no foot valves but mentions them in his text.[37] Francesco di Giorgio Martini, who, as Hollister-Short suggests, probably copied from Taccola, shows them. He also shows the grooves or 'cannelures' along the plungers that Taccola does not. He does not show the delivery spouts that are seen in Taccola's drawing.[38]

As far as I know, Hollister-Short is the first to attempt a description of how this pump might have worked.[39] He assumes an arrangement combining details from the two designs, with spout, groove and foot valve, and he assumes that the plunger is a close fit in its barrel.[40] His account is as follows. The plunger is raised; during the first part of the stroke, before the groove has reached the spout, water is sucked in through the foot valve. As the top of the groove passes the spout the suction is broken but the water remains in the barrel as the plunger is raised to the top of its stroke, its lower end clear of the water. The plunger is then dropped or at least brought down rapidly; it strikes the water, some of which is forced to rise up the groove and fall out of the spout.

There are difficulties with this account. First, the mechanism shown for raising the plungers is capable neither of being worked so as to bring them down violently nor of being detached to allow them to fall.[41] But worse than this, the level of the water in the barrel into which the plunger is to be thrust would, at best, be that of the lower end of the plunger when the top of the groove was opposite the spout.[42] By the time the plunger got down to the water, the outlet via the groove and delivery spout would be closed, and no pumping could be done. There is no way in which water could be drawn any higher with the arrangement outlined.[43]

To make the device work we must abandon trying to fill it by suction. The groove in the plunger might be longer, so as to communicate with the outlet through the full travel of the plunger, or there might be no groove, the plunger simply being an easy fit in the barrel. If the barrel were set with its lower end below the surface of the water, water would enter of its own accord as the plunger was raised, to the level of the surrounding water if the valve worked freely, as long as the air above it could escape freely.[44] When the plunger was lowered into the water in the barrel it would displace water that would pass up the groove or the annular space around the plunger, to be delivered at a spout or at the mouth of the barrel.[45] I illustrate the cycle in Figure 3.

Note that this way of working does not rely on bringing the plunger down fast, but the plunger must be heavy enough to sink deep enough in the water in the barrel or there must be means of pushing it down. These points are consistent with the arrangements drawn by both Taccola and Martini. There remains one problematic detail in Taccola's drawing, that he shows water issuing from the spouts of both of a pair of pumps at once; I can explain this only on the supposition that he wished to indicate that there were indeed two spouts, but there is no way in which water can issue from this pump, as reconstructed, while the plunger is being raised.

My reason for renaming this device a *pump with displacing plunger* should now be clear. Of course, the plunger in a force pump *displaces* the water that is delivered at each stroke, but then it is also the means of communicating an elevated pressure to the water; in this case the plunger does *no more* than displace it.

It is arguable that this device might represent a simplification of the force pump, suitable for use where the water is to be raised through only a moderate height and where it is not required that the water be discharged in a jet. Compared to the force pump, it calls for only one valve instead of two, and it does not require that the plunger be made and maintained a good fit in the barrel, a thing that remained quite hard to achieve, especially on a large scale, for some centuries to come. The supposition that this pump might indeed be closely related to the force pump is strengthened by the presence in Taccola of both types, appearing superficially very similar.[46]

Taccola does not draw grooves in the plungers and does not refer to such grooves in his notes. As I have mentioned above, my reconstruction of the device requires that the grooves, if present, must be long, but also that they are not necessary if the plungers are an easy fit in the barrels. So it is possible that in this respect Taccola's drawings are accurate.[47] One may ask, why then introduce the groove, as drawn by Martini? It might be that it was found convenient to make the plunger a less sloppy fit in the barrel for its better guidance, particularly when raised to its highest position (even though, as stated above, a close sliding fit would have been hard to achieve and to maintain).[48] This would be highly desirable if the plungers were relatively light and had to be pushed down, which Martini's mechanical arrangement seems to suggest. The groove would then present itself as an easily worked expedient to provide an adequate waterway.

That Martini does not show delivery spouts is odd, because without them or some other means of catching and directing what was pumped

Figure 3 Pump with displacing plunger (reconstruction after Taccola).

the water would tend to dribble uselessly back down the outside of the barrels. I can only explain this omission by appealing to Martini's frequent habit of exposing the hidden parts of machines in his drawings (such as the foot valves of pumps) and of not showing elements that almost certainly would have been present (such as fixed framing); perhaps he thought that the delivery spouts or launder heads at the top of the barrels could be taken for granted and that to draw them would be to clutter the illustration.

The next device that we must consider is named by Hollister-Short the 'hollow tube ram pump'.[49] Again I coin my own term to describe it, avoiding the word 'ram' not only because of the possibility of confusion with the *hydraulic ram* but also because it seems to me that the tube acts in no way like any sort of ram. I call it the *moving-tube lift pump*.

As Hollister-Short describes, this pump seems to work just like a conventional lift pump, with its foot and bucket valves; only the barrel is shortened, being no longer than is needed for the length of stroke of the 'bucket', while this 'bucket' is extended upwards as a tube to the level at which the water is to be delivered. The tube is raised by the action of a rotating cam lifting a tappet attached to it, and sinks back under gravity when the cam releases the tappet. The tube does not drop freely, but falls no faster than the passage of water (through the upper valve, with some leaking back through the foot valve and some leaking out between the moving tube and the barrel) will allow, and shock is no part of its working principle, just as it is no part of the working of the conventional lift pump. I illustrate the cycle in Figure 4.

A difficulty in interpreting Taccola's sketch is posed by his depiction of the water spouting from the falling tubes. This cannot happen, any more than that such water, if it were to be made to spout from upright tubes, might be relied on to fall conveniently, as drawn, to the side on which a trough is placed to receive it; there is nothing to push the water up but its displacement by the part of the moving tube that recedes into the outer barrel, and for the most part the water merely spills from the top of the tube as it sinks. A rational way to collect the delivery from this pump would be to turn over the end of each tube in a curve or an elbow, rather like that of Taccola's spouting water or as in my drawing; perhaps we have an indication here that Taccola drew the pump from memory and had forgotten this detail.

As with the force pump, the working of the moving-tube lift pump would be affected by leakage at the mouth of the barrel. If it were to be set above the level of the water to be raised, air leaking in would tend to spoil the suction, and probably the device would work well only if set with the barrel drowned. In other words, it shares with the force pump the disadvantage that only if made with superior workmanship can it raise water through a suction limb.

In itself, however, this pump embodies several elements of elegant and rational design. The extension of the reciprocating tube upwards to the delivery brings this moving part conveniently within reach of the driving shaft which is usually near the upper end of a water-raising apparatus.

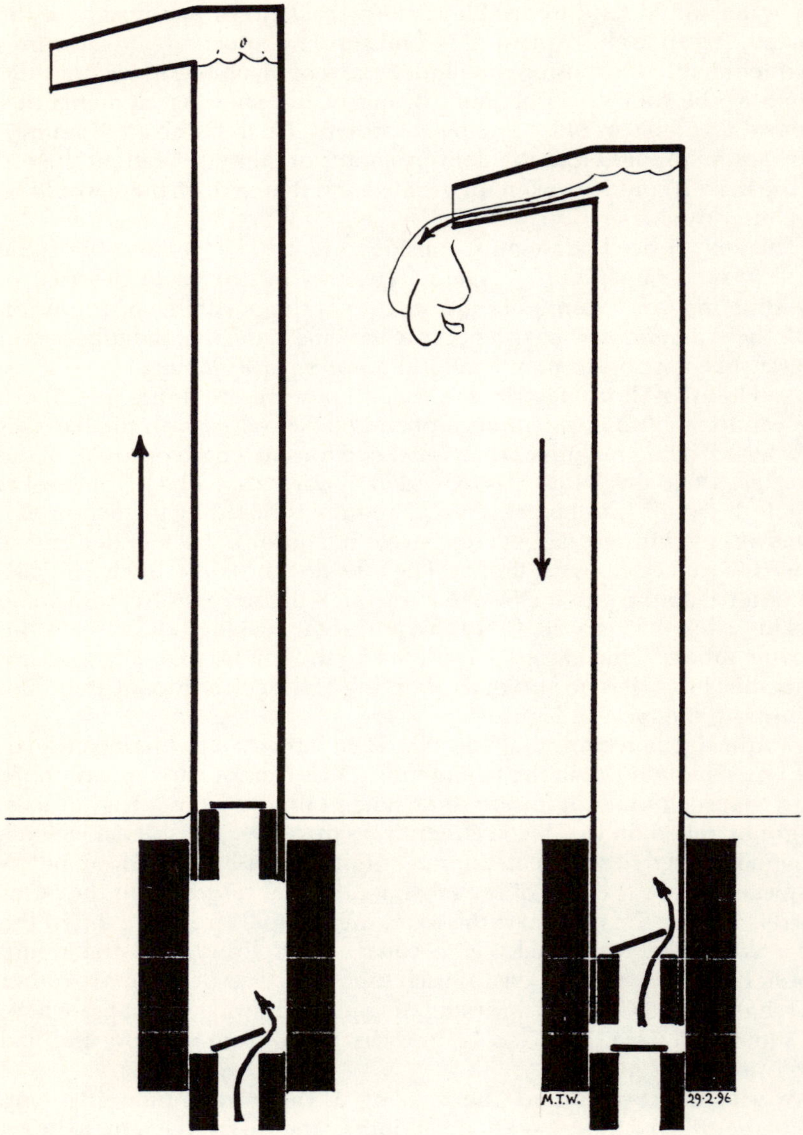

Figure 4 Moving-tube lift pump (reconstruction after Taccola).

Since the delivery pipe and some means of working the pump must both come to the top of the machine, there is a reduction in the number of parts of the device in making one component serve both ends. The pumps are shown grouped in pairs with a common tappet, so that the tubes would be prevented from rotating and the load on the tappet would be balanced laterally, and so be easier to constrain to move straight up and down. A detail in Taccola's drawing, not present in Hollister-Short's copy, is that these tappets are fitted with rollers so that the friction of the cams in raising them is diminished. Against all this, the weight of two tubes must be raised at each stroke (as well as the weight of the water that they contain), and even if they were made from thin sheet metal they would probably outweigh the apparent weight of the buckets and cords or rods of conventional lift pumps.[50] Taccola seems aware of this problem in specifying that the tubes should be of 'tin', presumably tinplate.[51]

COMPARISON AND CONCLUSION

The lift pump appears in Taccola's compilation along with a considerable number of other designs for water-raising devices. It is clear that a preoccupation with the problems of pumping had given rise to the play of a fertile inventive imagination, whether Taccola himself was the author of those designs or not.

Taccola includes sketches of what seems to be the force pump. The historical and archaeological records are silent on the point, but the type may have remained continuously in use since ancient times, and even if it was not used it may have remained known through the ancient literature; but Taccola gives little explicit detail in these drawings. In any case, if we compare these presumed sketches of force pumps with the lift pumps, the two types show no close affinity with one another.

I have suggested that the displacing plunger pump may be a simplification of the force pump, eliminating one valve and the most troublesome feature of the force pump, the need to make the plunger a close sliding fit in the barrel. Indeed, Taccola's sketches for this pump are very similar, in details of the plunger, for example, with the presumed force pumps mentioned above. The penalties of this design, as compared with the force pump, are that the machine is elongated vertically (although it may be made mainly of cheap materials such as wood), and that only a moderate lift can be achieved, because the pump cannot raise water higher than the top of its barrel. Moreover, the delivery at each stroke would be limited to the difference between the volume displaced by the plunger below the level of the surface in the lower reservoir and the volume of the space between the plunger and the barrel above this level and below the delivery; in other words, the delivery at each stroke is less than that got from a two-valved pump of comparable bore and stroke, but the pump could give good service for many applications. It contains one germ of the idea of the lift pump, that of raising the water within the extended barrel.

The moving-tube lift pump is much closer in its working to the lift pump as we know it, although its configuration is arguably quite close to that of the classical force pump with its rather short fixed barrel. In its use of a reciprocating pipe in which the water is raised, the upper end of which should, I suggest, be curved or provided with an elbow, it may owe its origin at least as much to ideas developed from the siphon,[52] which figures so prominently in the writings of Taccola and others (frequently misunderstood and in unworkable schemes!). By the use of two valves, it is, like the force pump, capable in principle of lifting water to any height provided that it is made to withstand the pressure and that a sufficient force can be applied to it. In practice, the need to manage the moving tube would limit the lift that could be achieved and the quantity raised at each stroke. The need to make it of unusual materials might perhaps have limited its use.

The application of all these designs except the lift pump is limited by the need to fix the pump at least partly below the surface of the water to be pumped: the displacing plunger pump, so that water will enter the barrel by gravity; the moving-tube pump, like the force pump, to prevent, or at least limit the amount of, air entering on the suction stroke. Both the displacing plunger pump and the moving-tube pump contain elements of the design of the lift pump and may be seen as intermediate between the force pump and the lift pump.

It is, however, doubtful whether we need to imagine the lift pump in direct competition with the force pump at this (apparently) early stage in its career. The two types seem almost to inhabit different worlds. Without ever explicitly stating a preference, Taccola seems to indicate their use in different applications. The force pump is shown raising water to a remote height, as in filling a siphon; the lift pump is shown raising water from below, as out of a well. The displacing plunger pump and the moving-tube pump are both shown raising large quantities of water through a limited height.

But the lift pump held greater potential for development than Taccola's other models: it was suitable for setting above the water in the lower reservoir because it is not so liable as the others to be troubled by air; and, especially when it came to building large pumps, the weight (and, since they are largely immersed, *a fortiori* the apparent weight) of reciprocating parts to be lifted was less than in the others. The reciprocating member was fully lubricated by water and could function well even if not a very close fit. Access to the foot valve for maintenance would have called for careful design, but at least it could be above the level of the standing water, and access to the valve and packing of the bucket, the other parts needing regular attention, was straightforward.

There is another advantage in setting the pump high and using suction, which may have been at least as important, especially with the limitations of the materials then available. In achieving a given overall lift, the higher one can make the suction limb, the lower the column of water above the foot valve needs to be. Thus the use of a high suction limb reduces the hydrostatic pressure in the pump. Moreover, reducing the height, and so the mass, of water above the bucket that runs back as the valves begin to close, only to be arrested suddenly as they seat, lessens the hydraulic shock delivered at each stroke to the fabric of the pump.[53]

It has been suggested that, since Taccola's drawings of the lift pump do not show a suction limb, the full potential of the pump had yet to be realized. It should be said that some drawings do not show the foot of the pump and so do not preclude the presence of a suction limb, and in this connection it is interesting that Taccola does illustrate, several times, one type of pump in which suction limbs are shown, even drawing one as a separate detail with an explicit annotation in one case.[54] Two large pairs of bellows of conventional cuneate shape are set on a horizontal frame above water, with the suction limbs reaching down below them. A man, hanging by his hands from an arrangement like a swingle-tree, transfers his weight from the back of one bellows to the other. Cords connect the ends of the swingle-tree to the upper boards of the bellows, so that as one is depressed the other is drawn up. Water is shown issuing from the nozzles of both bellows, which, as in another case discussed earlier, I take to be a convention of the drawing; in reality, water would issue from the two nozzles alternately. For the arrangement to work, there must have been non-return delivery valves in the nozzles, which may be what Taccola intends to show in another drawing.[55] It is perhaps worth pointing out that this type of pump, being relatively immune from the fault of having air leaking in during the suction stroke, was a good choice for use with a suction limb, a further example of sound rational design. Mention is made in the appendix of another pump illustrated and described by Taccola which may (but this is less certain) also have had a suction limb.

We may conclude that the lift pump had been developed by the time that Taccola completed his *de Ingeneis* (that is, before 1433) and that the suction limb also existed by that date, and had been used with a pump, perhaps with two pumps, which were particularly suited to sucking. It is therefore quite possible that the lift pump was also used with a suction limb by this date, although we have no evidence that it was. We do know from Agricola that the lift pump with suction limb was being well exploited a century later in Saxony.

There is no reason to suppose that Taccola was the originator of the designs that he drew.[56] The presence among his sketches of rather inept schemes for the application to pump valves of draw-strings and floats tends rather to suggest that he was not a very competent designer of hydraulic machines; or, at least, they suggest that he was not aware of the superiority of the better designs that he also drew and perhaps that he did not understand fully how they functioned.[57]

Taccola's sketches of the displacing plunger pump and of the moving-tube pump do show the use of elements of the lift pump in designs that are arguably close to the force pump; in the first the water is raised up the extended barrel, while in the second the delivery valve is placed in the reciprocating element. But I cannot trace a clear evolutionary path from the force pump through these two to the lift pump; the displacing plunger pump seems very close to the force pump, but the moving-tube pump, which seems closer to the lift pump in the way in which it functions, does not have the extended barrel.

It is perhaps safer to conclude that these two pumps may represent parallel developments of the force pump, and that the lift pump may have been the product of either line or of another, third line of development;

or that elements from one line of development were grafted on to a design originating in another line of development.

In any case, we see with hindsight that the design of the lift pump, with its reciprocating bucket always working drowned and so lubricated by water and proof against the destruction of the suction by air, even when the pump is above the level of the water to be raised, offered greater potential for development than the other designs of reciprocating pump of Taccola's time. And the subsequent history of the lift pump, from Agricola to the present day, is proof that this design was remarkably successful.

APPENDIX

The Chain Pump

In the chain pump water is raised up a valveless pipe by a series of bodies which fit the pipe reasonably closely and which are connected together flexibly into an endless loop. The loop runs over a drum at the top which is rotated to give motion to the loop, and may run over another drum at the bottom for guidance.

As with the lift pump, Taccola gives us the first known illustrations of this pump.[58] In one place he characterizes it as coming from the Tartars, which tempts us to think of this pump as a technology transferred from Eastern Asia, where a comparable device in which rectangular pallets on a moving loop of chain, drawing water up an inclined open-topped flume, is known and appears to be ancient.[59]

Taccola shows, and describes, a chain of wooden mushroom-shaped elements. More usually in later pumps these are replaced by stuffed leather balls or more or less rigid discs separated by several plain links of chain. Agricola shows the chain pump performing heavy duty in mine pumping,[60] and Bate illustrates versions worked by a hand winch, a tide wheel and a windmill, as well as mentioning a horse-driven example in London raising water from the Thames for domestic supply.[61] The chain pump continued in use to recent times, two important advantages of the design being that there are no valves to go wrong and that it lends itself simply to being driven by a hand winch or by any rotative source of power.

The Diaphragm Pump

Taccola gives several sketches of this pump and a rather enigmatic description.[62] It has been called the 'sack pump' because Taccola describes a flexible component 'fastened like the sack for straining the dregs out of wine'. Unfortunately this is less than clear to us because we do not know how the wine-strainer with which Taccola was familiar was arranged. What we see in the sketches is a membrane pulled up into the shape of a tent by a cord fastened to the middle. In some of the sketches there is a characteristic fitting, presumably of metal, to which the cord is tied.

Earlier commentators have taken the tent shape to be the upper part

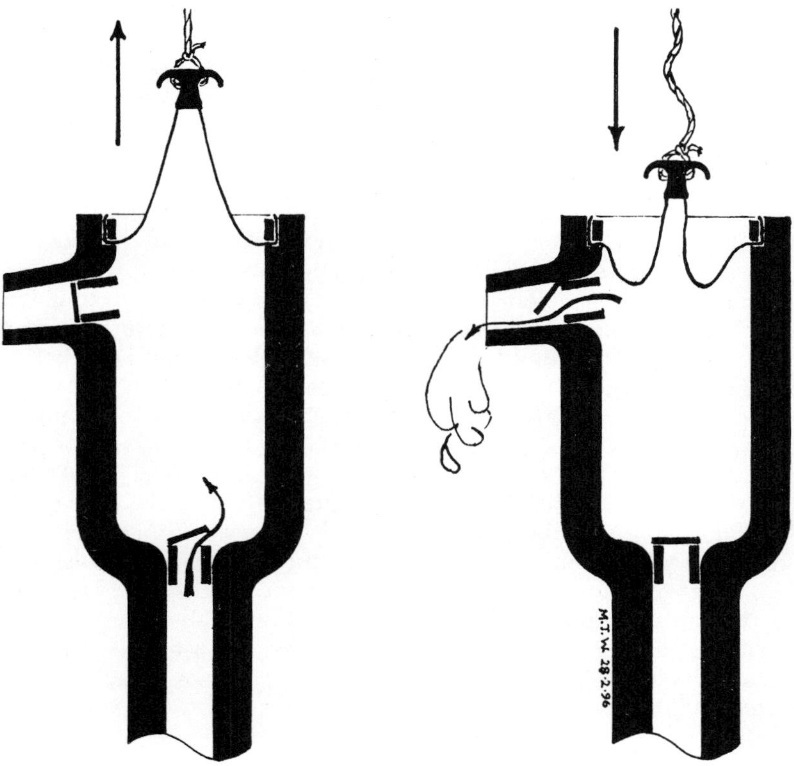

Figure 5 Diaphragm pump (reconstruction after Taccola).

of a bag that is worked up and down in the pump-barrel like the bucket of a lift pump; they imagine that it contains some filling that allows it to let water past as it descends but not as it is lifted, but I cannot see how this is to be done unless the bag is subtle enough in its design to have called for further comment from Taccola.

It seems to me more probable that we have here the earliest appearance of the *diaphragm pump*, in which a flexible membrane is worked in and out to vary the volume of a vessel to which it is attached, in effect as a substitute for the barrel and plunger of the force pump. So I envisage the tent-shaped 'sack' as a circular piece of fabric, perhaps leather, attached by its edge to the cylinder: an arrangement that could well be likened to a form of filter. Taccola draws the body of the pump as a short cylinder of a rather large diameter, which suits this arrangement. To the cylinder is fitted a foot valve, perhaps with a suction limb, and a delivery spout that must also be fitted with a valve. I illustrate this reconstruction in Figure 5.

Taccola describes a number of lengths of terracotta pipe, suggesting that the pump is raising water from some depth. This, and the detail of his sketch showing a marked reduction in diameter below the uppermost part of the pump, strongly suggest that the pump has a suction limb.[63]

Note that, as with the bellows pump described above, my reconstruction of this pump is not liable to be seriously affected by the leakage in of air to spoil the suction, as long as the delivery valve shuts adequately and the diaphragm is adequately fixed and is not ruptured. Terracotta would seem to be a poor choice for the manufacture of a barrel in which anything, let alone a bag of cloth or leather, is to slide. Taccola's apparent enthusiasm for this pump might perhaps reflect the reaction of a man used to the problems of maintaining a sliding seal to a pump from which this troublesome element was eliminated. If I am right in my interpretation of the drawing, the design was indeed a good one because the diaphragm pump continues in use today.

There is one sketch that at first sight shows this type of pump but which, if it does, presents us with a difficulty.[64] Taccola shows water discharged not at a tubular spout some way down the barrel but at a lip at the top like that of a jug. This of course means that the water is getting above the 'sack', if that is what Taccola has drawn. Conceivably there might be a valve, even merely a flap of leather opening upwards, in the 'sack'; this is easy to imagine if the lift, and so the suction, is not too great. But there is no note, and it may be that in this sketch we see again the displacing plunger pump, the plunger carelessly drawn somewhat tapering and shaded so that we might mistake it for a 'sack'. The ingenious reader is at liberty to choose between these explanations or to develop his or her own.

Notes and References

1. I prefer to avoid the term 'suction' in naming the pump for two reasons: there are those who may object that suction does not exist, although in common with many others I use the term to mean the depression of the pressure in an enclosed space below that outside, which effects the entry of a fluid into that space; more seriously, there is no reason why, in principle, the force pump may not suck just as the lift pump does, although reasons why it may not suck as well in practice will emerge from what follows.

2. G.J. Hollister-Short, 'On the Origins of the Suction Lift Pump', *Hist. Tech.*, 1993, 15: 57–95. Dr Hollister-Short's notes include bibliographic detail of the most important previous contributions on this subject. The basis of Dr Hollister-Short's paper is a consideration of designs for pumps illustrated by Taccola. Readers to whom detail and context are important should be aware that his figures are *redrawn* from the facsimile sources that he cites (and which I give below, note 32).

3. It is tempting to speculate that the term reveals the shape of the component in early pumps; certainly the bucket in Taccola's lift pump (see below) is deep, but the valve is shown at the top, not the bottom. But I am not aware of any parallel between the term in English and in any other language, as there is, for example, in the case of *cock* (English) and *Hahn* (German).

4. Thus the Torricellian vacuum is not a true vacuum, but always contains vapour of the liquid substance used in the barometer at its equilibrium vapour pressure, even if air is carefully excluded; but the vacuum in a mercury barometer is better than that in a water barometer because the equilibrium vapour pressure of mercury is so much lower than that of water at any temperature.

5. That is, at right-angles to the tangent to the surface.

6. A shock delivered to the water at any point will be propagated as a system of pressure

waves, to be dissipated by the viscosity of the water and by communication to the container and by movement of the water if it is free to move. Shock can be used to pump water, as in the *hydraulic ram* invented by J. Whitehurst in 1772, but usually it is to be avoided as being very destructive to the delicate parts of the system such as the valves of a pump. The name of the hydraulic ram makes it unfortunate that force pumps have sometimes been called ram pumps, and their plungers, rams; and for the same reason I avoid the names using the term 'ram' used by Hollister-Short (see note 2) to describe two types of pump illustrated by Taccola.

7. I illustrate my argument by reference to a pump with a plunger, working vertically, corresponding to the earliest type known; but what follows applies equally well to force pumps with pistons and with the plunger or piston working in any direction.

8. The reader may object that if the pump is drowned suction is not required to make it fill; but water flows only where there is a 'pressure gradient', a local difference in pressure. In this case the pressure inside is lowered by the withdrawal of the plunger, and that depression of the internal pressure may be called suction.

9. It is necessary to suppose that the proportions of the pump are such as to develop enough suction at each stroke to draw water into the barrel. This performance will often depend on keeping within bounds the quantity of air in the barrel, and it may be necessary to suppose that the pump is *primed* (that is, filled with water) to begin with and that its design is such that much of the air entering the barrel, by whatever means, will be expelled on the delivery stroke.

10. The extent to which the actual performance of a pump falls short of what a simple calculation (neglecting such sources of loss) would suggest is called the *slippage* and is usually expressed as a percentage of the calculated figure. For a simple pump the slippage may well be 10 per cent or higher.

11. In the Roman bronze pumps that survive, the plungers have always been found to be relatively thin-walled castings, open at the bottom, as I have drawn them (see note 28). In use these voids would remain nearly filled with air, which has been remarked on by some commentators as advantageous in providing some elastic cushioning against shock, comparable to the air vessel applied to more modern pumps. In reality, any such effect would be small and of doubtful value; shock would be a significant problem in such a pump only if it were pumping against a high head such that much water in the rising main ran back at the end of the delivery stroke. To protect the delivery valve, which would be most at risk, any air vessel must be fitted to the side of the delivery valve away from the pump. Rather, these voids may be seen as detracting from the pumps' performance during the suction stroke. Some writers have suggested that they may have been filled with wooden blocks (which could have had leather packing washers nailed to the bottoms, as in some of the surviving Roman wooden pumps), but I am not aware that any evidence for this practice or for closing the void with a disc has been reported, and it is not mentioned in the contemporary literature. The design of these bronze plungers makes very good sense both in economizing material and in making the ruling section of the casting fairly uniform.

12. A slackness of fit that allows only a moderate leakage of water may be sufficient to allow a quite serious leakage of air. One can conceive of a force pump fitted with a reservoir surrounding the mouth of the barrel into which water is run from the rising main, so that any leakage inward past the plunger would be only the ingress of some of that water; but I am not aware that this expedient has ever been used in ordinary circumstances.

13. The relative quantities of water delivered on the two strokes may be controlled by adjusting the proportion of the cross-section of the rod, at the level of the water surface in the pump, to that of the bore of the barrel. This is seen, for example, in Smeaton's apparatus for determining the efficiencies of water wheels, in which Smeaton arranges for equal deliveries on the two strokes. See John Smeaton, 'An Experimental Enquiry Concerning the Natural Powers of Water and Wind to Turn Mills', *Phil. Trans.*, 1759, LI; the arrangement is shown clearly in Plate V.

14. It was common to set pumps obliquely in mines when the excavation itself was narrow and not vertical.

15. This is why it is helpful to prime such a pump by pouring water into the barrel.

16. The joints pose a problem, particularly due to the cycles of stress as the pump is worked, whatever materials are used for the pipes. The pipes themselves, if of metal, may

present few problems but over a very long period it was common to make pumps and their associated pipework largely of wood.

17. The 'suction limb' is a length of pipe leading from below the surface of the water to be raised up to the pump, through which water is sucked into the pump. A further development of the lift pump, the jack-head pump, will not be considered here. In this, the top of the barrel is closed by a cover with a gland through which the bucket is worked by a rod; a side branch from near the top of the barrel carries the delivery to some height above the pump itself, as in the force pump. It will be appreciated that for the purposes of the present discussion its working is exactly that of the lift pump.

18. Georg Bauer, *alias* Georgius Agricola (the elder), *De Re Metallica,* libri duodecim. (Basle, 1556); translated into English from the first Latin edition by H.C. Hoover and L.H. Hoover (London, 1912).

19. 'I have omitted all those things which I have not myself seen, or have not read or heard of from persons upon whom I can rely. That which I have neither seen, nor carefully considered after reading or hearing of, I have not written about' (translation by Hoover and Hoover, xxx, xxxi).

20. It is perhaps worth mentioning that in Agricola's woodcuts the arrangements are often compressed vertically to make the fullest use of the page.

21. See, for example, T. Ewbank, *Hydraulic and Other Machines for Raising Water* ... (London, 1842).

22. See Hollister-Short, *op. cit.* (2).

23. The reconstruction in Guido Ucelli, *Le Navi di Nemi* (Rome, 1950), 185, Figure 200, of lift pumps from remains found in the ships in Lake Nemi has rightly been discredited by S. Shapiro, 'The Origin of the Suction Pump', *Tech. Cult.*, 1964, 5: 566–74.

24. J. Bate, *The Mysteries of Nature and Art* (London), 57–60. I have consulted the second edition of 1635, which is said to contain new material; I do not know whether his account of the London Bridge waterworks appears in the first edition of 1634.

25. N.A.F. Smith, *Man and Water* (London, 1975), 101.

26. H.W. Dickinson, *Sir Samuel Morland, Diplomat and Inventor, 1625–1695,* Newcomen Society Extra Publication no. 6 (Cambridge, 1970), 57, 83. Morland's way of arranging the packing seems, however, to have been defective.

27. J.P. Oleson, *Greek and Roman Mechanical Water-lifting Devices: the History of a Technology* (Toronto, 1984). Oleson discusses both the literature and the archaeological evidence in detail. T. Schiøler, 'Bronze Roman Piston Pumps', *Hist. Tech.*, 1980, 5; also T. Schiøler and J.A. Garcia-Diego, 'Bronze Roman Piston Pumps', Proceedings of the Symposium *Ancient Technology* (Helsinki, Tekniikan Museo, 1990). The latter is a revised version of the former with some new photographs. The authors discuss technical aspects of the bronze pumps in considerable detail.

28. Ibn al-Razzāz al-Jazarī, *The Book of Knowledge of Ingenious Mechanical Devices* ... , translated and annotated by D.R. Hill (Dordrecht, 1974), 186–9, 266. The original book was completed in 1204 or 1206. Different interpretations of the mechanism driving the pump are possible; Hill suggested a pair of spur gears but Schiøler (private communication: TS to MTW, 1995) more plausibly suggests that what was intended was the right-angle *sāqia* gear. In either case, the ensemble of the pump itself remains recognizable, the two barrels merely turned with their mouths towards one another. Nevertheless, Hill preferred, for reasons connected with the terms in which the description is couched, to regard it as not directly derived from the pumps of Hellenistic antiquity.

29. Oleson, *op. cit.* (27).

30. Granger, in his bibliography to his edition in the Loeb Classical Library series, cites, for example, 17 copies thought to range in date from the tenth to the fifteenth century, seven of them from the fifteenth century: Marcus Pollio Vitruvius, ed. and trans. F. Granger, *de Architectura,* 2 vols (1931), *Vol. 1,* xxxii–xxxiii. The first printed edition of Vitruvius was that of Sulpitius, *c.*1486.

31. What follows is based on published sources for Taccola's work, as follows: J.H. Beck (ed.), *Mariano di Jacopo detto il Taccola: Liber Tertius de Ingeneis* ... (Milan, 1969); G. Scaglia, *Mariano Taccola: De Machinis, the Engineering Treatise of 1449* ..., Vols I and II, (Wiesbaden, 1971); F.D. Prager and G. Scaglia, *Mariano Taccola and his book De Ingeneis* (Cambridge, MA and London, 1972); G. Scaglia, F.D. Prager and U. Montag, *Mariano Taccola: De Ingeneis* ...

Books I and II . . . , Vols I and II (Wiesbaden, 1984). In making reference to illustrations in the above, I give both the folio number of the manuscript and the page number of the printed book. In the case of Beck, the manuscript is reproduced in facsimile on pp. 41–136, but these are not paginated; there is both Arabic numeration of the folios, which is consecutive, but not always easy to read, and Roman numeration which does not run consecutively. In making reference to these pages in Beck I give both the (Arabic) folio number and, in parentheses, the page number. A useful commentary by Lon Shelby, comparing the first three of these books, is to be found in *Tech. Cult.*, 1975, 16: 466–75. Of course, this predates the publication of the fourth book cited above.

32. The chain pump uses a number of 'pistons' in rapid succession to draw water up a valveless pipe; it probably has no part in the present discussion beyond being one among several water-raising schemes. It is, however, important in its own right and by Agricola's time it was performing heavy duty. The device to which I refer as a diaphragm pump is what Hollister-Short, *op. cit.* (2), calls a sack pump, but he and I differ as to how it may have worked. I do not think that its working provides as close a parallel to that of the lift pump as he supposes; therefore I leave it out of the present discussion. I give an account of both these pumps in the appendix.

33. Scaglia, Prager and Montag, *op. cit.* (31), *Vol. II*, p. 164 (f. 82v.); redrawn as Figure 3 in Hollister-Short, *op. cit.* (2).

34. The conditions would be: first, that the barrel was set low in the water and that the bucket would dip to some depth below where the water would rise to in the barrel naturally; second, that the bucket would rise near enough to the top of the barrel that some of the water so trapped above it would pour out.

35. This drawing from *de Ingeneis* III appears in Beck, *op. cit.* (31), f. 3r (p. 45), with the description on f. 3v (p. 46) transcribed on p. 138, and in Prager and Scaglia, *op. cit.* (31), Plate 25, p. 74, the text transcribed and translated on p. 75.

36. Beck, *op. cit.* (31), f. 8 (pp. 53, 54); redrawn as Figure 5 in Hollister-Short, *op. cit.* (2).

37. Beck, *op. cit.* (31), pp. 140, 141.

38. The version of this pump by Martini to which I refer is redrawn as Figure 7 in Hollister-Short, *op. cit.* (2).

39. *Op. cit.* (2).

40. This assumption should not be allowed to pass without comment. To achieve and maintain, without a resilient packing, a sliding fit between the plunger and the barrel, on the scale that seems to be envisaged, that was sufficiently close to allow the pump to suck would have been hard to achieve. It may, however, have been achieved, at least on a small scale, because elsewhere Taccola draws what seem to be force pumps, indistinguishable from the type presently under discussion except that the delivery comes from the *bottom* of the barrel. The foot of the plunger of the force pump might have been furnished with a packing such as that found in Roman wooden force pumps (see Oleson, *op. cit.*, 27). But it also remains possible that these drawings represent pumps that were not actually made or did not actually work.

41. Taccola shows the two plungers of a pair of pumps connected to balance beams, to the other ends of which are attached ropes that are wound in opposite senses on the barrel of a windlass; Martini shows the beams linked to vertical racks that engage on opposite sides of a pinion. In either case, one plunger is made to descend as the other is raised, and both move at the same speed.

42. In reality the water level in the barrel would be lower, for the reasons outlined in discussing the behaviour of pumps; see above.

43. Dr Hollister-Short tells me (private communication, 1996) that the late Dr Donald R. Hill seemed prepared to accept his reconstruction with some (unspecified) qualifications; but unless what Hill envisaged was, in effect, my different interpretation which follows, then I cannot see what he had in mind.

44. This does not contradict the point made above that water only flows where there is a pressure gradient; in this case, the difference is between the *hydrostatic* pressures near the lower end of the barrel owing to the different depths of water inside and outside. Considerations of 'suction' are irrelevant because the space in the barrel is not enclosed.

45. This mode of working has also been suggested by Dr Lynn Willies, in a private communication to Dr Hollister-Short.

46. See above (40). A clear example of a force pump is seen in Scaglia, Prager and Montag, *op. cit.* (31), *Vol. II*, p. 126 (f. 63v). Taccola annotated this drawing to indicate that the wooden plunger, which was to be raised by a rope and to drop under its own weight, was to be surmounted by a cap of stone or metal.

47. The device seen in Scaglia, Prager and Montag, *op. cit.* (31), *Vol. II*, p. 116 (f. 58v), in which water wells up all round a square plunger in a square barrel, may then be seen as just another version of the same device. Presumably both barrel and plunger of this pump were to be made by fastening boards together, the latter to be suitably weighted, perhaps with water or stone. In this case, the delivery is caught by a gutter surrounding the barrel.

48. Note also that Martini shows the plungers with upward extensions passing through eyes in the fixed framing, a clear indication of his concern to guide the travel of the plungers.

49. Beck, *op. cit.* (31), f. 19 (pp. 75, 76); redrawn as Figure 6 in Hollister-Short, *op. cit.* (2).

50. In the lift pump, the apparent weight of the bucket and rod is less than its weight in air because it is immersed.

51. 'Metallo, quod dicitur stagno'. Solid tin, an expensive, soft and dense metal, would make no sense in this application. Tinplate, sheet-iron coated (by fusion) with tin to prevent it rusting, would make quite good sense, except that the tin would tend to be worn off where the components rubbed. Tinplate seems to have been manufactured at least as early as 1340 in Bavaria; see W.E. Minchinton, *The British Tinplate Industry* (Oxford, 1957), 2. In footnote 3 on p. 1, Minchinton refers to the following works on the early development of the tinplate industry: F. Hessel, *Die Zinnblechhandelsgesellschaft zu Amberg* (1914); F.M. Ress, *Geschichte und wirtschaftliche Bedeutung der oberpfalzischen Eisenindustrie* (1950); O. Johannsen, *Geschichte des Eisens*, 3rd edition (1953). It appears still to be true that the subject has not received a full treatment in English. Taccola's is the earliest reference that I know to tinplate, if that is what is intended, other than as a material for domestic and similar utensils. There is, however, an uncertainty of nomenclature, at least in English, where the word *latten*, which may mean tinplate, is often read, I suppose by comparison with the French *laiton*, to mean brass; the problem is compounded by modern writers who affect the word latten to describe copper alloys.

52. Σίφων seems to have been the technical term for the Hellenistic force pump, first found in Hero (see Oleson, *op. cit.* 27), but I use 'siphon' here in the modern sense of a tube in the form of a ∩ in which a flow may be set up by having the two openings at different heights.

53. In this analysis I disagree with Hollister-Short (*op. cit.*) in the conclusion that he draws from a passage in R. D'Acres, *The Elements of Water-Drawing* or *The Art of Water-Drawing* (London, 1659 and 1660 respectively). (The variant titles are found on the different title pages of two issues of the same printed text. The book was reprinted (under the second title) with an introduction by Rhys Jenkins as Newcomen Society Extra Publication no. 2 (Cambridge, 1930).) The passage reads: 'The deeper you let fall your *Bucket*, the easier it will go, and not so soon subject to be out of order; it is better to lift a foot of water than suck or draw a foot.' This passage appears in the context of the arrangement of pumps to lift water through a height about as great as that theoretically possible by suction or more. As I read it, D'Acres refers not to the possibility of damage to the pump but to the problems owing to cavitation and leakage of air that I discussed above under 'Mechanics', which become much more insistent as the theoretical limit of the suction lift is approached, and which cause the pumping engineer to opt as a compromise for a suction lift comfortably less than the limit.

54. Scaglia, Prager and Montag, *op. cit.* (32), *Vol. II*, p. 62 (f. 31v) show the suction limb separately; another pump of this type, or another drawing of the same pump, is on p. 212 (f. 105v).

55. Scaglia, Prager and Montag, *op. cit.* (31), *Vol. II* p. 125 (f. 63r) This 'scrap view' may, however, relate to an adjacent drawing that seems to show a different arrangement using a bellows, perhaps to fill a siphon, and as far as I can make it out the valve opens inwards, in the opposite direction to the way required for the present purpose. Such valves are needed

to prevent the bellows from simply drawing air back through the spouts instead of sucking up water.

56. There is perhaps a stronger case for suggesting that the more obviously fanciful arrangements are Taccola's own. For example, the displacing plunger pump and the moving-tube lift pump are both included in *de Ingeneis* III in the context of working water-mills; but it is the elements, in this case the pumps themselves, rather than the uses to which they are put that concern us more closely.

57. That is, if I have understood those drawings correctly. Examples are found in Scaglia, Prager and Montag, *op. cit.* (31), *Vol. II*, p. 120 (f. 60v) and p. 144 (f. 72v).

58. Scaglia, Prager and Montag, *op. cit.* (31), *Vol. II*, p. 160 (f. 80v); and Beck, *op. cit.* (31), ff. 4r, 4v (pp. 47, 48) and p. 138.

59. J. Needham *et al.*, *Science and Civilization in China, Volume 4* (Mechanical Technology), Part II (Cambridge, 1965), 339ff.

60. Agricola, *op. cit.* (18).

61. Bate, *op. cit.* (24), 13, 14, 57–61, 78, 79.

62. Examples may be found in Scaglia, Prager and Montag, *op. cit.* (31), *Vol. II*, p. 169 (f. 85v); and Scaglia, *op. cit.* (31), p. 61 (f. 40r). The latter includes the description.

63. I attempt to illustrate this feature, which does not show in Hollister-Short's version of Taccola's drawing, in my Figure 5.

64. Scaglia, Prager and Montag, *op. cit.* (31), *Vol. II*, p. 212 (f. 105v).

Acknowledgements

The problem of the origin of the lift pump came to the fore in lectures by Dr N. A. F. Smith and Dr G. J. Hollister-Short that I attended nearly twenty years ago. I owe to these two friends a continuing debt of gratitude for the stimulating exchanges that have over the years contributed materially to keeping my brain active. But I would not have looked so attentively at Taccola's drawings, and so this chapter could not have been written, but for Dr Hollister-Short's paper on the same subject; he has been most generous with his time in discussing of points arising out of that paper and in sharing with me correspondence that he has received. Both Dr Hollister-Short and Dr Smith have read drafts of this chapter and offered helpful comments. Naturally, I myself bear the responsibility for what I have written.

Discoveries, Inventions and Industrial Revolutions: On the Varying Contributions of Technologies and Institutions from an International Historical Perspective

IAN INKSTER

INTRODUCTION

This presentation ranges rather widely, from the conditions of discovery in eighteenth-century Britain to the character of industrialization in modern East Asia. Between these extremes I also briefly consider the nature of industrialization in late developers of the nineteenth century, such as Russia, Hungary, Poland and Japan. The prime focus is on the role of discovery and invention in cases of successful industrial revolution. My overall claim is that 'discovery' is a constructed category, a real phenomenon dependent upon social and institutional resources as much as upon great men and flashes of insight. Second, I suggest that as time goes by, as the historian passes from the late eighteenth-century industrialization of Europe towards the fractured, heavy industrial programmes of late nineteenth-century Russia or Hungary, into the world of the later twentieth century, then so too discovery becomes of far less importance in the industrial success of nations. New products, processes and even entire new industries (e.g. microelectronics or biotechnology) are of significance in the continuing development of mature industrial systems in Europe, North America and Asia. But the breakthroughs which we label as 'industrial revolutions' appear to be less and less dependent upon 'discoveries' within their own boundaries and more and more dependent upon their own equally creative institutional responses (innovations) to a forever changing external world.

In the first main historical section of this paper I consider the first industrial revolution (that of late eighteenth-century Britain) in the light of its most famous technological case, James Watt's separate condenser of 1765 onwards. Whether following Gilbert Ryle or, more recently, Simon Schaffer, this seems to fit the bill for a 'discovery'.[1] That is, Watt's separate condenser breakthrough was individual, isolated, recognizable and reliant upon precursors, and produced identifiable results/impacts, which were of specific and temporal occurrence. There was a 'moment of discovery' on Glasgow Green, Edison's 'burst' of intuition, either associated with or producing a sudden insight or closure of a problem. When describing the 'context' of Watt's discovery, it would be quite feasible to do so in terms of J. P. Guilford's or Liam Hudson's conceptions of divergent and flexible thought processes.[2] Notably, recent work suggests that modern scientists and engineers have a strongish tendency towards convergence rather than divergence in mental make up.[3]

The possible dynamic links between discoveries and inventions in mature industrial systems on the one hand and institutional innovation in 'follower' systems on the other has yet to be properly explored by historians or others. Bearing in mind only the theme of discovery and invention, we can suggest three groupings of industrial revolutions. In the British and perhaps other earlier cases, discovery and invention loomed large and was arguably: (a) socially and institutionally resourced; (b) forged within a local context of a hybrid technical culture which had long benefited from transfers-in from other nations; (c) recognized as significant through the activity of a host of more mundane, incremental improvements; and (d) variously dependent on market forces which prompted usage, e.g. the price of coal influenced the choice of Watt engines over improved Newcomen engines in the early nineteenth century.

In the later industrial revolutions of the years prior to 1914–18, discovery and invention was of lesser importance and the ability to transfer established advanced technology from elsewhere was of the essence. Perhaps because the technology transferred into late developers was, on the whole, already fairly standardized in the older industrial states (explaining the continuing importance of transfer through trade), the institutional innovations required to sustain the borrowing process tended to be in the direction of emulation. That is, along with technologies, institutional models were also transferred-in as part of the wider modernization process. However, it might be noticed that within this borrowed and shared paradigm there were a myriad normal variations and choices, and the case of Japan, which moved from a British-style institutional model towards a Prussian one in the 1880s and 1890s, is a very salutary one.

Third, later twentieth-century industrial revolutions are similarly dependent on transferred techniques and institutional innovations. But in this group, much of the transferred technique (especially from the early 1970s onwards) is new and barely utilized within the originating systems. Indeed, the Japanese and South Korean cases have recently shown that effective commercial operations may be reached within the nation of late industrial revolution *prior* to such a level being attained in the nation of origin, a phenomenon which has prompted unwise accusations that East Asian industrialization has caused the fall of the Atlantic-based economic system. Furthermore, because of the shift in the technological imperatives of new and growing industrial sectors (especially microelectronics), and the pervasive character of such industries, successful late industrialisation is now occurring on the basis of institutional innovations or refinements which are not merely emulations of the market or other institutions of the nations of origin. Indeed, there is a repeated claim that East Asia succeeds just because the direction of its institutional development has been other than that of America and Europe, perhaps towards that of Japan or the wider, so-called 'Confucian' style of management, enterprise culture and social relations.

INTERPRETATIONS

Happily, there are one or two 'industrial revolutions' going on right now. Large parts of East Asia and arguably one or two nations of South America, perhaps also Turkey, appear to be undergoing precisely that combination of economic growth, structural change and institutional development from low levels of per capita income which we, as historians, associate with the much-maligned term *industrial revolution.*[4] Modern China provides us with the expected measures: the tremendous increase in the share of industrial output in national income, the technological and organizational transformation within agriculture, the very marked reform of a wide range of market and non-market institutions during 1978–85 and so on. There are other outstanding features of the contemporary Chinese industrial revolution: the centrality of *technology transfers* from other, more sophisticated places; the large part played by *foreign* investments; the role for a new 'openness', itself instigated by reforms towards the market determined in most part by government itself.[5] Indeed, the detailed regional pattern of industrial development, from the coast and south spreading inland and north, is clearly determined by the dynamically interacting forces of proximity to foreign techniques and finance and to overseas trade and cultural influence.[6]

So, here we have a contemporary story of industrial revolution without much in the way of *discovery* and *invention.* Of course, modern Guangdong is awash with newness, with awakening; the system rewards the bright-eyed. And clearly enough, a 'discovery', as a cognitive event, is yet a discovery, however many individuals situated geographically, industrially and, dare I say, cognitively 'elsewhere' may regard it with disdain as mundane. Again, an *organizational* solution to a production problem is still a solution in one place, even if formulated quite conventionally in a more sophisticated system 'elsewhere'. As we shall see, use of terms such as 'discovery and invention' does require some notion of site and agency. Nevertheless, when all is said and done, we may reasonably entertain the feeling that to adjust and to adapt, more clearly to define and utilize, is not quite the same thing as to 'discover' – where are the equivalent Benjamin Huntsmans, James Watts, Jethro Tulls, Edmund Cartwrights and Henry Maudslays in very late industrializations?

These are surely names to conjure. Do they not immediately bring to mind that first industrial revolution; Britain somewhere in the eighteenth century when, surely, something happened?

Although at times admitting that recent and often very sophisticated and hard-won 'estimates of industrial production are approximate', and that the even more complex estimates of overall growth are 'controlled conjectures', some recent scholarly work has tended to emphasize continuity of change by rescuing earlier growth phases while dampening the rate of change in later stages.[7] Nevertheless, Nick Crafts and Knick Harley have both agreed quite recently (1992) that, for instance, 'industrial innovations – did create a genuine *Industrial Revolution* reflected in changes in Britain's economic and social structure'.[8]

Of course, this is not as such a story of 'discovery and invention', but that story has in the past been quite convincing and is now being approached from novel angles, albeit involving pathways that are often lit by quite traditional and almost conventional understandings, reaching back to David Hume in the early eighteenth century or T. B. Macaulay in the nineteenth. So, in a very recent essay, the leading economist and economic historian Donald McCloskey has concluded from his enquiry into the reasons for Britain's early industrial lead that 'It is no accident . . . that the nations where speech was free to a fault were the first in economic achievement: Holland, Scotland, England, Belgium and the United States.'[9] Anglocentric readers might well note the astute ordering of the nations.

This perhaps surprising contemporary echo of *enlightenment* rationality and the later Whigs, of John Stuart Mill in 1848, with the emphasis on freedom, varied agency and competitive emulation, on social institutions as *causative* of industrial revolution, is increasingly the favoured stance of a variety of historians as they turn away from the national accounting procedures of the 1970s and 1980s. Appealing for a return to the historian's more reliant emphasis on 'multiple causation', David Landes has judged that in late eighteenth-century England 'something happened' and 'that something was essentially technology'. And England's lead developed out of a relative ease of social interaction, a consequent generation of a vital and exceptional skill environment and the resulting 'precocious recourse to water and steam power in mining and industry'.[10] From quite another direction, the French meta-historian Fernand Braudel emphasizes the great noise of new technologies *throughout* eighteenth-century Europe – gearing, transmission, mining and rolling machinery. Braudel believes that technological history is potentially history as such ('the history of technology is that of human history') and is therefore forced to address what to him is a very significant problem: why did all this clamour of technique take so long to make any impact on actual economic production and why did the later productive combinations of coke and coal exploiting techniques emerge in Britain of all

European places?[11] I am perhaps being unfair, but it seems that Braudel's own answer is not so very clear. Britain was different in being saddled with less 'social inertia' but blessed by a high degree of craft specialization; that is, the Braudel conclusion follows lines also drawn by McCloskey and Landes. In reference to a point that Joel Mokyr has recently addressed more technically and compellingly, Braudel also singles out the lesser resistance of workers as of some importance in the British case.[12]

So, in summary, we have some fresh agreement on the significance of technological change, on a degree of discontinuity in the history of techniques and on the importance of some form of social and institutional resourcing of technological change in the early British industrial revolution. And although dressed anew and often very elegantly, much of this would not have come as a great surprise to many earlier commentators and analysts.

The patent data do bear this out quite well. The gross patent statistics show a kink in output around the mid-1700s. However, if we systematically exclude from the figures the patents for medicines, drugs, dental apparatus, musical instruments, decorative design and interiors and furnishings, then a kink appears around the 1770s and 1780s, of a sort perhaps more notable than can be found for the conventional measures of change, whether that change be considered as *explanan* or as *explanandum.*[13] Nor is it at all difficult to locate a contemporary feeling of change, disruption or uncertainty focused on what were certainly regarded as the new technologies. This is quite well known, but I offer you the Reverend William Turner, dissenting minister of the city of Newcastle, writing in the year 1797. Turner was no outright critic of new machinery, indeed he spent great efforts on promoting machinery and knowledge of machinery throughout Newcastle and its district for most of his lifetime. But having just toured the textile districts of Yorkshire and Lancashire, Turner could not refrain from the following comments:

> When the owners of a great factory take three hundred children from the overseers of St Giles, and bring them down in covered wagons, never to see, or hear of, or care for their parents more, what can we call such vehicles, but slave vessels upon wheels or what such factories but what I hear one of them emphatically called, a receptacle for white negroes?[14]

We shall return to William Turner of Newcastle at a later point. For the rest of this paper I wish to centre on two themes only. The first concerns the character and external resourcing of 'discovery and invention' in the first industrial revolution, especially as exemplified in James Watt and the separate condenser. Attempts not only to generate but also to multiply 'genius' (as exhibited in some of the reform talk relating to Britain's *present* educational system) might be of little importance in initiating significant technology change, which within any system might be more the outcome of novel institutions, social relations and career opportunities.

Our second theme is the seeming decreasing function of 'discovery and invention' in subsequent industrial revolutions. Late industrializers in Eastern Europe, Russia and Japan were characterized by large-scale processes of technology transfer in which the ability to adjust and absorb became of greater importance than the proclivity to discover and invent. Our overall theme, then, is that the world is now some complex and unforeseen outcome of technical and institutional novelties, in which the importance of the first in initiating industrial change decreases as time goes by, just as the importance of the second (institutions) increases. At the present time the locus of new industrializations may rest in those nations or regions best able to invent institutions. Just as older industrializers may remain the centres of new scientific and technical discoveries and inventions, so too may they become less and less able to emulate or learn from a complexity of new institutional formulations, 'discovered and invented' in Pacific economies, where Confucian rather than Protestant cultures display an ability to generate historically exceptional economic and industrial performances.[15]

JAMES WATT AND ALL THAT

Let us turn to James Watt, the eighteenth century and notions of 'discovery and invention'. In a major speech to the Third Annual Meeting of the British Association in 1833, William Whewell (1794–1866), later Vice-Chancellor of the University of Cambridge, emphatically concluded that 'we cannot create, we cannot even direct, the powers of discovery', a point now echoed by the economist Donald McCloskey, who argues that true creativity and discovery cannot be routinized, or they would not have value.[16] But what discovery? Clearly, the discovery of benzene rings is not equivalent to the discovery of the separate condenser, which is not similar to the discovery of scotchgard, which does not seem at all identical to the discovery of the *concerto grosso* or to the discovery of the White Nile; and none of these seem to be the same as the discovery, by the Earl of Sandwich, of Omai, that native of Raiatea, refugee sheltering in Tahiti, who had been placed in the dubious care of Sir Joseph Banks and subsequently suffered a typically tyrannous late eighteenth-century season of favour among the worldly yet credulous *cognoscenti* of Enlightenment Europe. If we focus by using the term 'discovery *and* invention', then Omai thankfully drops out, as does the White Nile, but we are still left with, for instance, both benzene rings and the *concerto grosso*, that inventive seventeenth-century juxtaposition of the orchestra and the *concertino*. After all, notions of contrast, tension and juxtaposition are at the very core of many psychological accounts of both creativity and invention, including that offered by Arthur Koestler.[17]

We might be able to separate out the abstractions (for instance, of science) from the seemingly concrete or material reality of machines and equipment. For instance, as demonstrated nicely by Simon Schaffer in a recent essay, there is surely a sense in which scientific 'discoveries' must be *constructed* or 'made up' through the artful dodging or deliberate actions of key players; that is, others must be convinced, and at least part of this 'conviction' must come from some form of intellectual or cognitive alteration, some process of persuasion involving change of mind.[18] In this sense, among several others, scientific discoveries are socially resourced. On the other hand, in the case of *productive technique*, discovery is more basely observable and laudable and its significance in terms of commercial value lies to a great extent beyond the values placed on it by other, competing and perhaps very jealous engineers or artisans.

Some years after his statement of 1833 (in 1857), William Whewell was prepared to divide *scientific* breakthroughs into moments of genius-discovery and processes of humble incrementalism, a division made subtle and very useful in the work of Thomas Kuhn and his followers.[19] But the division and the relationship between the great and the mundane in the history of *technology* might be more appropriate than in the case of *science*. The distinction between breakthroughs and increments in technological history might depend less on authoritative, communal adjudication based on seemingly intellectual grounds and much more on observation of and attention to the sequencing and crowding effects characteristic of events which are normally physically recorded (e.g. in a layout, or machine or design) – patent history shows the crowding of incremental technique improvements around already recognized technological breakthroughs.[20] Thereby, Watt's separate condenser of 1765 becomes a very significant 'discovery and invention' while J. T. Desaguliers's use of the injection-cock in the Savery engine or James Pickard's adaptation of Newcomen's atmospheric engine to rotary motion (1780 patent) are more clearly *incremental*; items among the many recorded remains of a host of followership events occurring in the eighteenth century and beyond.[21]

Yet we may not rest sanguine in any way. The full and useful Watt engine required far more than the separate condenser, but embodied additional developments such as steam jacketing, double acting cylinders and parallel motion, which do perhaps seem more incremental and certainly relied for their applications on skills shared with the incrementalists too. Furthermore, we have to accept, I think, that both great advances and more routine steps and adjustments shared a common *external resourcing*, from the skills, institutions and social relations which surrounded them or, more accurately, their agents. Thus the famous historian of Europe, H. A. L. Fisher, referred to the 'atmosphere of the age' and Donald Cardwell, Britain's best-known historian of technology, writes of the skills and freedoms which promoted not only discovery and incrementalism as defined above but all those new efficiencies arising from improved engineering standards; in the case of steam engines, better boilers, furnaces, cylinders, lagging and so on.[22]

The trouble is that one historian's *context* becomes another historian's *text*.[23] Can we focus more on Watt around 1765 and emerge with some argument which helps us to understand why there were some Watt-like 'discoveries and inventions' in later eighteenth-century Britain, but also why not all artisans and engineers were Watt-like. An intuitively obvious and potentially strong argument might refer to Watt-like innovative abilities, insights and so on and their very rare occurrence in the eighteenth century or at any other time. But we might also wish to make reference to the specific local and social *siting* of the inventive individual.

A goodly number of commentators would find fault with my formulation here, especially in centring on James Watt.

It is seemingly clear enough that the sort of Newcomen atmospheric pumping engine that Watt was faced with in the early 1760s required the sudden water-cooling of the cylinder in order to condense steam, which in turn created a vacuum beneath the piston, allowing the pressure of the atmosphere to provide the working stroke of the piston downwards; the piston rose again, not so much by steam expansion as by the counterweight of the pump rod. Watt's first insight was in realizing that a continually heated cylinder would improve the engine's performance, followed then by his solution that the *separate condenser* (wherein steam rushed into a separate smaller vessel evacuated by an air pump) would satisfy the dual requirement: thus the Watt illumination during the famous stroll on Glasgow Green in May 1765, described by himself and others, but at a later date.

> In the Green of Glasgow, and when about half-way between the Herd's house, and Arn's Well, my thoughts having been *naturally turned* to the experiments I had been engaged in for saving heat in the cylinder, at that part of the road *the idea occurred to me*, that, as steam was an elastic vapour which would expand, and rush into a previously exhausted space; and that, if I were to produce a vacuum in a separate vessel and open a communication between the steam in the cylinder and the exhausted vessel, such would be the consequence.[24]

There is no real argument against the idea that Watt's early solution somehow resulted from specific endowments of the sites in which he worked and this approach is well enough chronicled by scholars – the problem set by the need to work on the Newcomen model, Watt's knowledge of the then trajectory of the incremental improvements surrounding him and the subsequent external resourcing through skilled labour, materials and tools, financial support and institutional protection through the patent system to the year 1800. Within this richly resourced environment emerged the accumulative Watt technology already noted, as well as the governor and the later sun and planet gearing for rotative power.

But does the heroic element now disappear? Does discovery and invention become not necessary but rather a mere reflection of the inevitable outcome of some wider (still British!) social and technical *context*? Does context take over?

In their recent well-known account *Multiple Discovery*, David Lamb and Susan Easton argue that Watt simply 'participated in a particularly important stage in the evolution of the steam engine' and that in the absence of Joseph Black's conception of latent heat 'Watt could not have developed his steam engine'. Thus the authors emphasize that Watt 'took' his 'problem' to Joseph Black, who 'explained' the phenomenon of latent heat, which 'revealed' to Watt the reason for heat/energy loss, which he then 'applied to the design of a superior steam engine with a separate condenser and a steam jacketed cylinder to convert the steam back to water'.[25] Watt was, essentially, caught within a particular trajectory of mechanical improvement, and Lamb and Easton's 'explanation from context' seems to remove the heroic or highly individualized contribution. A particular story about 'context' is used here not only to 'explain' Watt but, in a sense, to explain him away, to convert a breakthrough into an incremental change. The separate condenser and much else has been routinized. But can we still give a 'social' story which stops short of translating Watt into a mere Mertonian accelerator?

We can begin by briefly questioning two key elements in this reduction of Watt. First, do we *know* that Joseph Black's latent heat was necessary to the initial phase of problem definition? Way back in 1952, Fleming first made the claim for Watt's independent 'discovery' of the principle of latent heat, as against the view of the humble artisan, situated at some site 'below' that of the 'scientist', being duly informed of the principles upon which he was actually working.[26] Schofield's formulation is something in between. Schofield notes a transition in Watt's site of activity during the 1750s which was equivalent to a transition in his *character*, a 'transformation from a relatively unschooled mechanic to an educated scientist'.[27] This may be too abrupt and anachronistic. When John Robinson first met Watt at the latter's workshop, he 'saw a workman and expected no more', though he changed his mind soon enough as he experienced the experimental Watt, his turn to the steam engine during 1758–63, the escalation of his interest during 1762 and so on.[28] Watt's later statement of Black's influence at this stage was that 'the correct modes of reasoning and of making experiments of which he set me the examples, certainly conduced very much to facilitate the progress of my inventions'.[29]

What are we to make of these terms: conduce, facilitate, inventions? At the time when Watt was more or less forced to reflect upon the 1760s in defence of his patents (in the 1790s), he was apt to emphasize an experimental past: thus in a letter to James Hutton of December 1795, 'I do not believe even in Mechanicks without experiment to which test I wish to bring all theoretical opinions if possible'.[30] After noting Watt's experiments on the mechanical properties of steam, measurement of elasticity and so on, Schofield judges that it was 'in consequence of these experiments that Watt rediscovered the phenomenon which Black explained to him by the theory of latent heat'. Furthermore, 'All of the central ideas to be incorporated in Watt's various steam engines for the remainder of his life were essentially worked out in his mind by the spring of 1765'.[31]

Second, why is it unproblematic that Watt moved (after 'months of pondering') from the clearer definition of the problem to its engineering solution? What evidence from the scholarly work of Fleming, Dickinson, Law, Cardwell, Musson or Tann is there that another mechanician was about to do the same? The evidence appears to be that both Newcomen and Smeaton early recognized the degree and significance of heat loss without inventing the separate condenser. In the work of the historians we uncover a Watt soaked in the formal and tacit skills of his grandfather and father, pictures of Isaac Newton on the wall, a reading of Desaguliers, all-but routine inspections of Savery and Newcomen models, an individual embedded in the *Zeitgeist*. But too much emphasis here may miss the point. We also need to address the *non-typicality* of Watt's location. *Zeitgeist* and typicality arguments merely evoke the question: if the *Zeitgeist* commands and directs and if historians and contextualists can push it back to the ancients (as did – in this case – the historian of science A. R. Hall and the sinologist Joseph Needham[32]) then why did Watt-type interventions take so long? *Zeitgeists* are not sufficient; nor are 'contexts' that go back so far. Watt's specific historical locale permitted (and perhaps induced) the drawing upon resources from the artisanal culture *and* the savant nexus *and* the commercial support structure which culminated in his partnership with Matthew Boulton in Birmingham.

Watt's capacity to come up with the separate condenser was, then, sited within discrete, very particular components of changing technological culture, components that by no means merged together for the host of other mechanicians facing mechanical puzzles – the Glasgow university workshop, the shops and workshops of the Salt Market and (by 1763) the Trongate, the savant nexus of the Lunar Society, the routines and challenges of the travelling surveyor (e.g. the ditch canal between the Forth and the Clyde) and the Soho connection forged in the summer of 1768. Watt's 1765 'discovery' appears to have been 'made up' during 1768 and 1769 (the time of the first patenting) and this involved reformulation of Watt's contribution by himself, Boulton, Small and others in a fairly intensive and self-conscious manner. There is a great deal of non-typicality about this, and in the immediately subsequent years (1770–2) of important engine experiments where the focus was on the solution of very practical mechanical matters (e.g. piston packing).

So Fisher's general 'atmosphere of the age' or even the *technological Zeitgeist* might be a 'context of explanation' in terms of both the resourcing of technical innovation and the acceptance and recognition of technical innovation. But perhaps only a firm acknowledgment of variations and pockets within the *Zeitgeist*, particular locations, helps to explain the important 'discoveries and inventions'. A reason for judging the issue in such a way derives perhaps from the dualism involved in the idea of *expertise* or *embeddedness*. Clearly, expertise might be useful in searching and focusing in any puzzle-solving situation, but too great an expertise and too great a degree of site stability might produce a tendency towards convergence, towards myopia. In a normative theory of reasoning, intuitions become judgements about which specific inferences are acceptable and which are not – thus the Master chess player does not search the myriad thousands of possibilities but selects 'immediately' perhaps just three or four. It is this, indeed, which distinguishes experts from ordinary folk.[33] British 'artisanal culture' (*Zeitgeist*) *attuned* the expert Watt to the nature of the problem. His local contexts, between social worlds, might then have acted as the essential social resourcing for his *disturbance*, for the reduction of a myopia arising from the power of past and present technique conventions, by challenging established technological categories, providing analogues and juxtapositions, injecting intuitions which broached the existing technical discourse.

It was the habitual encounters of later eighteenth-century artisanal culture which produced the many incremental solutions, but perhaps Watt-like discoveries demanded something more in the way of external resourcing, something local and particular. This allows the possibility, I think, that eighteenth-century incrementalists differed from Watt not solely in talent, in persistence and the propensity absolutely to exhaust a technique niche, but also in either not being sufficiently expert to be able to perform a heuristic search or to be so embedded in their own expertise as to be perfectly capable of normal puzzle-solving but less likely to produce breakthrough discoveries. Watt was both embedded and disturbed. In summary, we may have here a heroic Watt within a heroic site! And is it the system which could produce this sort of combination over and over again which secured the first industrial revolution?

CATCHING UP AND TAKING OVER

The world of incrementalism had several functions. In what A. P. Usher called the frequent 'acts of skill', techniques transferred out from Britain and became the material implements of a complex round of late industrialization, of industrial revolutions centred in Germany, or Russia or Japan prior to about 1914.[34] While Usher and others emphasized the importance of skills and personnel in the transfer of technology from advanced nations to relatively backward economies, the Novocastrian William Turner stressed the changing character of machinery itself, and recognized transfer as a potential problem for Britain way back in 1797. Writing from Newcastle, Turner staunchly defended the advancement of machinery in England, particularly in the textile industries, even though it 'has been attended, in the outset, with much riot and disorder'. But he also saw a long-term disadvantage to Britain's interests arising from increases in quality, falls in price and standardization. Prior to this, best technique had leaked out of (and, of course, into) England mostly via individual agency, especially through the migration – often under contract – of skilled English labour to nations of lesser technological sophistication. According to Turner, this had hitherto been *ad hoc*, at least reducible through government legislation (e.g. restrictions on skill migration which lasted – decreasingly effectively – into the 1840s) and of limited impact. Now, however, other nations could catch up with English industry through the importation of advanced and trustworthy machinery directly, especially as machines, unlike English labour, 'have no prejudices to conquer, no relatives or friends to leave, no old habits to break, no new language to learn, no rooted attachment to their native soil'.[35] This was to be well borne out soon enough: witness the diffusion overseas of British high-pressure steam engines in the years following the Napoleonic Wars.

However, the technology transfer phenomenon, although partial enough, stretched around the globe. The argument here is that the keys to the spread of industrial revolutions prior to 1914 lay decreasingly with discoveries and inventions, moments of intuition or genius, and increasingly with elite reaction and institutional innovation; with responsiveness and the ability to sustain long-term technological projects through periods of commercial loss and enormous *political* strain.

By the later nineteenth century, both the agency and the siting of 'discovery and invention' had been revolutionized within leader nations such as Britain, the USA, Germany and France. Free floating engineers and entrepreneurs abounded and inventions yet emerged from the world of the apprenticed artisan or mechanic, but increasingly patent attorneys, engineers' societies and associations, government bureaus of standards and plant industries, agriculture and chemistry, programmes of focused research in individual private enterprises and engineering colleges and polytechnics dominated the process of 'discovery and invention' in the advanced systems.[36]

Such relatively sophisticated sites of endeavour and negotiation served to move information and technique around among the winner economies and combined technical and legal expertise with the tacit understandings and advantages gained from clubbability and new types of 'habitual encounter'. In Britain it was the legal beavers and engineers of Southampton Buildings, Chancery Lane or Fleet Street and Gray's Inn; in the USA the Corcoran and the Norris Buildings in Washington harboured an equivalent mix, while in New York the Inventors' Institute, Electrical Exchange Buildings and the many specialist offices on Broadway and on Nassau Street housed the clusters of engineering and legalistic initiative. In Paris the specialist engineers and inventors' societies dominated, just as in Berlin a vast amount of patenting was organized by the Central Patent Bank, the Patent Anwalte (Verlag) and the Centralblatt für die Deutsch Metal-Industrie Spandauerstrasse. Within such sites there was big business; the French associations handled thousands of 'discoveries and inventions', while the single legal and engineering firm of H. and W. Pataky of Berlin and Prague had processed over 12,000 patents by the end of the nineteenth century.[37]

It was in and through such sites that technological competition and transfer took place among the small group of industrializing nations. In just the short period 1886 to 1900 the six largest organic chemical firms in Germany lodged no fewer than 948 coal-tar chemical patents in the *British* patent system, 'discoveries and inventions' which had emerged from the employment within these elite enterprises of 845 trained chemists and engineers. As to sheer scale, between 1877 and 1894, 343,581 patents were applied for in the USA and 266,544 in Britain. As to focus, of all patents lodged in Germany between 1877 and 1904 over 12,000 related directly to the chemical industries and over one-quarter of these were lodged by engineers, firms and legal agents resident overseas.[38]

So the face of 'discovery and invention' was highly 'made-up' and Janus-like; for every transfer of information and equipment into an industrial system there was both a potential benefit and a potential loss to individual enterprises, entrepreneurs or engineers. It was a system which created the major patterns of the twentieth century as we knew them until around the 1960s.

Among those systems which emerged as 'winners', the invention of institutions seems to have been most drastic within the later members, such as Germany, the USA and Japan. This was well recognized within the British Association itself. A. G. Green belaboured the Association in 1901 on the subject of Germany's transition from dependency status to a centre of scientific and technical excellence:

> There can be no question that the growth in Germany of a highly scientific industry of large and far-reaching proportions has had an enormous effect in encouraging and stimulating scientific culture and scientific research in *all* branches of knowledge. It has reacted with beneficial effect upon the universities, and has tended to promote scientific thought throughout the land. By its demonstration of the practical importance of purely theoretical conceptions it has had a far-reaching effect on the intellectual life of the nation.[39]

In what became a well-known speech at the time, the President of the British Association in 1903 addressed his audience on the subject of 'The influence of brain power in history' and asked, 'shall we follow Japan and thoroughly prepare by "intellectual" effort for the industrial struggle that lies before us?'[40]

The query has a certain modern ring to it, does it not? New, increasingly complex, interrelated and sophisticated institutions were, then, visualized as the real requirements for industrial health, and perhaps national survival. By June 1905, Japan's victory over Russia yielded the first evidence of the significance of late industrialization in the East. The leading British journal *Nature* was perhaps more solemn than the President of the British Association when it proclaimed that the 'lesson which our educationalists and statesmen have to learn from Japan is that the life of a modern nation requires to be organised on scientific lines in *all* its departments ... it must be consciously used for the promotion of national welfare'.[41]

In this sort of context, what of *very late* industrialization, the nations met with as we move eastwards through Hungary and Poland to Russia and beyond? We might highlight this theme by briefly considering the huge mining–metallurgy–engineering complexes grafted on to the economies of southern Hungary, southern Poland and Russia (around the Black Sea in the area bounded by Odessa, Kharkov and Rostov) in the years prior to World War I. In these separate attempts at industrial revolution from conditions of extreme backwardness we may spot features more akin to present-day economic modernization in China than to the eighteenth-century British industrialization or even to that of nineteenth-century North America – the massive injections of foreign capital, joint ventures, large importations of advanced equipment all linking very late industrialization to the continuing progress of the 'winner economies', including those of Germany and even Austria, and perhaps above all the initiating and controlling role of the nationalist state at all levels of intervention, from tariffs and direct investment (especially in mining and railways), through technology search and application, to labour regulation and the speedy erection of training institutions.[42] The technological story here is of great interest, but has two features of particular relevance.

Each industrial complex possessed its own trajectory, made up of an accelerated transition from relatively simple transfers of Bessemer steel technology or Austrian tensile testing to extremely convoluted and interrelated accumulations and adaptations to original blueprints. Thus, in South Hungary the Reschitza iron and steel works was founded on the basis of relatively simple modernized ironworks taken over from the government in the 1860s. By the 1880s Reschitza combined French, Belgian, British, German and Austrian technologies in a massive complexity, involving some 80,000 workers in several locations, incorporating key Hungarian-manufactured components (e.g. the first Hungarian locomotive from 1873) and increasingly sophisticated crucible steel plants, hydraulic and electrical haulage systems, adapted Siemens-Martin technologies and so on. Such a massive complex employed skilled workers from Britain, Germany, Italy and France as well as from Hungary, Romania and Bulgaria.[43] By the turn of the twentieth century, the technological character of very late industrialization bore little resemblance to its *own* beginnings, never mind to those of, say, eighteenth-century Britain.

Second, in those late nineteenth-century developments we can also discern a certain transition from a narrow *statist* dominance centred on directly strategic products (guns and butter), to a phase of greater *private* sector activity in which originally transferred techniques were applied to a far greater range of final products incorporating entirely new industries and locations. Thus, by the mid-1880s the core metallurgy of south-west Poland and Warsaw had spawned a large manufacturing enterprise, producing equipment for new agricultural implement factories, breweries, flax, saw and sugar mills, tramways and large civil engineering schemes. The reasons why such great projects ultimately failed to act as the bases of more generalized industrial revolutions (e.g. despite Walt Rostow's claim that Hungary reached an economic 'take off' late in the nineteenth century, in the year 1900 only 13 per cent of the active population of Hungary was employed full-time in industry, compared, say, to 38 per cent in the then united Germany[44]), lie beyond the scope of this discussion, and are at any rate much debated.[45] Clearly, however, industrial failure in such areas was strongly associated with political and institutional breakdown, with the inability of the social system to survive the strains induced by such enormously enclavist developments.[46]

The power of novel *institutions* over that of new *technologies* is perhaps no better illustrated than in the great contrast between late industrialization in Eastern Europe and Russia and late industrialization in Meiji Japan (1868–1912). This introduces a very long story, but in essence Japanese industrialization resulted from an extended and close association between technology transfer and institutional transformation across all aspects of Japanese society, embracing such phenomena as the establishment by government of educational and training institutions (the single largest item of normal expenditure to the 1890s), official searching, testing and selecting of advanced foreign techniques, the foundation by government of model factories and mills, the expenditure of huge amounts on measures of both direct political control and cultural manipulation, the disestablishment and dispersal of traditional, aristocratic interests and groupings, all of this concentrated in a short phase of frantic activity between 1868 and 1885.[47] The Japanese industrial drive of these years was not associated with a *national* cultural transformation, but with what we might term a conscious process of 'cultural engineering' controlled by an industrializing elite; Japan was not, then, Westernized, but it did break the back of late industrialization in the years prior to 1912.[48] Yet this was not, on the whole, a tale of 'discovery and invention' in any conventional manner, but instead one of varied institutional innovations and adaptations, moving from an original focus on Anglo-American institutions to a later one (1890s) centred on Prussia, an economy of later development.[49] One or two institutional features do stand out and may be summarized.

First, the government's dual attention to manufacturing industry and to physical infrastructure (especially railways, docks, harbours and lighthouses) was serviced by a pragmatic *intelligence system* composed of Japanese students and officials in combination with foreign employees, teachers and traders. Perhaps surprisingly, in the light of modern images of Japan, foreign employees were not utilized as mere technical go-betweens, but were often closely induced into the Japanese decision-making system – thus Dr Paul Mayet's profound influence as savant and highly paid adviser to the Financial Section of the Council of State.[50]

Second, the *institutional interface* between Japan and the industrial West was not fixed or, indeed, defined, but constantly in flux, centrally incorporating not treaties but the Westerners of the treaty ports, voluntary associations in which foreigners and the Japanese elite mixed (some of these nominally illegal, but in fact more than tolerated), 'habitual encounters' between Western teachers and their Japanese college and university students, and a great system of co-mingling in regional and industrial exhibitions, fairs and so on reaching down to levels below that of the provincial prefecture.[51]

Third, we should note the extremely high status of the Japanese technology transfer agents, incorporating the Imperial Court itself, as well as the Cabinet and the leadership of such emerging Westernized companies as Mitsui and Mitsubishi.[52]

A last institutional feature worth emphasizing here is the measurable emergence of private sector initiatives during the mid-1880s and beyond, surely following upon:

- the social control measures of the 1870s (to about 1882);
- the demonstration effects of government models;
- the relative financial calm in the aftermath of Matsukata's deflation;
- the increased supply of educated workers and of modern-skilled labour from the arsenals and dockyards;
- the cost-reducing effects of greatly improved transport and haulage infrastructure.[53]

This, of course, is a tedious way of suggesting that a multitude of expensive and often inefficient state interventions in the field of technology transfer and settlement prior to 1885 created the conditions for private enterprise innovations after the mid-1880s.

CONCLUSIONS

May we now draw a few threads together? In 1889 Rudyard Kipling referred to Meiji Japan as merely 'a button-hook appanage of America'. Things have changed – apart from holding a vast amount of American assets, Japan (followed, you may note, by China) today holds the largest current account surplus with the USA. Industrial revolutions made a difference, and history was not all a smooth curve, nor was industrial revolution at all divorced from basic socio-institutional contexts. And little was predictable. If Britain would not have been picked as a 'winner' in 1650, then neither was North America in 1750, Japan in 1850 or China in 1950. But (at least potentially) there seems to be a more particular story about technology and institutions which is worth the telling.

The industrial revolution in England appears to have been associated with three types of technical change: 'discoveries and inventions' of the Watt-type; incrementalism of the adaptation and accumulative type; and what we might name Cardwell-type improvements in standards, e.g. in materials, finishing, specifications. All of these were of importance in eighteenth- and nineteenth-century Britain and I suggest that all were more or less products of a complex of institutions and social relations within which 'habitual encounters' of the normal kind generated accumulative technical improvements, while local pockets of variance promoted breakthroughs or deviations from the norm.

In contrast, most cases of later industrial revolution drew upon 'discoveries and inventions' established and elaborated elsewhere, and the success of later, often state-instigated, programmes depended more on the establishment of institutions which encouraged adaptation to the physical and organizational requirements of advanced techniques. Work by economists on contemporary Indonesia, Yugoslavia and China suggests that sustained industrial growth in conditions of extreme initial relative economic backwardness depends enormously upon increasing the *technical efficiency* of already transferred advanced machinery, and that this results from both improved standards and institutional flexibility.[54] But the initiation of the modernized industrialization process is associated with statist interventions in the form of newly institutionalized hierarchies and centralism, bureaucracy and legalism, cabal-like relations between government and favoured private sector agents (as may be discovered in Russia and Japan in the 1890s, or South Korea and Indonesia in the 1960s and 1970s); indeed, an *institutional context* quite other than that found within early industrializers. Furthermore, it may have been (and be) the ability of the political system to maintain this structure in the face of dissent, outside pressure, inflation (or its equivalent) and gross inefficiencies which determined the success of late industrialization, rather than variable degrees of technical capability as such. In the latter respect, the Russians were initially far more capable than the Japanese.

We may conclude with David Landes that, viewed from the perspective of the relations between technological and institutional change, industrial revolutions were not 'gentle bumps on a pre-existing trend' but 'wrenching and compelling forces for change'.[55]

Notes and References

The present chapter is a somewhat formalized but substantially unchanged version of a lecture delivered under the title of 'Industrial revolutions: are discoveries and inventions necessary?' to the Annual Festival of the British Association for the Advancement of Science at Newcastle upon Tyne, 11 September 1995. The overall thematic title of the festival was Discovery and Invention. I would like to thank Frank James and Alec Chrystal for inviting me to the meeting, and the British Association and the Joseph Nelson Group for full financial support of my trip from Sydney, Australia. I would also like to give thanks to Gerry Martin, Penelope Gouk and the members and guests of the Achievement Project, Oxford (1990–5) for incubating some of the thoughts which are expressed in this chapter.

1. G. Ryle, *The Concept of Mind* (London, 1949); S. Schaffer, 'Making up Discovery', in M.A. Boden (ed.), *Dimensions of Creativity* (Cambridge, MA, 1994), 13–51.

2. J.P. Guilford, 'Traits of Creativity', in H.H. Anderson (ed.), *Creativity and Its Cultivation* (New York, 1959), 141–61; L. Hudson, *Contrary Imaginations* (London, 1966).

3. R.S. Mansfield and T.V. Busse, *The Psychology of Creativity and Discovery* (Chicago, 1981); R.J. Sternberg (ed.), *Creativity, Current Psychological Perspectives* (Cambridge, 1988); R.W. Weisberg, *Creativity, Beyond the Myth of Genius* (New York, 1992).

4. For discussions of the term of particular interest to historians of technology, see A. Paulinyi, 'Revolution and Technology', in R. Porter and M. Teich (eds), *Revolution in History* (Cambridge, 1986) and several papers in J. Mokyr (ed.), *The Economics of the Industrial Revolution* (Totowa, NJ, 1985).

5. Wang Hongmo *et al.*, *China 1949–1989* (Zhengzhou, 1989); Liu Guoguang *et al.*, *The 1980s: China's Economic Reform and Development* (Beijing, 1991); N.R. Lardy, *Foreign Trade and Economic Reform in China* (New York, 1992).

6. Led generally by privately owned enterprises (with collective ahead of state enterprise growth) by the early 1990s significant per capita GNP growth was occurring in even the most inland of regions.

7. C. Knick Harley, 'Reassessing the Industrial Revolution: A Macro View', in J. C. Mokyr (ed.), *The British Industrial Revolution* (Boulder, CO, 1993), 171–226, p. 176. For well-written review of the issue generally, see D. Cannadine, 'The Present and the Past in the English Industrial Revolution, 1880–1980', *Past and Present*, 1984, 103: 131–72.

8. N.F.R. Crafts and C.K. Harley, 'Output Growth and the Industrial Revolution: A Restatement of the Crafts-Harley View', *Economic History Review*, 1992, 45: 703–30, p. 713.

9. D.N. McCloskey, '1066 and a Wave of Gadgets: The Achievements of British Growth', in P. Gouk (ed.), *Wellsprings of Achievement: Cultural and Economic Dynamics in Early Modern England and Japan* (London, 1995), 114–32, p. 124.

10. D. Landes, 'The Fable of the Dead Horse; or, The Industrial Revolution Revisited', in Mokyr, *op. cit.* (7), 132–170, pp. 157, 162.

11. F. Braudel, *Civilisation and Capitalism 15thC–18thC. I, The Structure of Everyday Life* (London, 1981), especially 370–428.

12. J. Mokyr, 'Technological Inertia in Economic History', *The Journal of Economic History*, 1992, 52: 264–79.

13. See Chapter 3 of I. Inkster, *Science and Technology in History: An Approach to Industrial Development* (London, 1991) and the stylized R.J. Sullivan, 'The Revolution of Ideas: Widespread Patenting and Invention During the English Industrial Revolution', *Journal of Economic History*, 1990, 50: 349–63.

14. *The Monthly Magazine*, IV, July–December 1797 (London, 1798), pp. 255–6. For a good account of William Turner and Newcastle science, see A.D. Orange, 'Rational Dissent and Provincial Science: William Turner and the Newcastle Literary and Philosophical Society', in I. Inkster and J. Morrell (eds), *Metropolis and Province, Science in British Culture 1780–1850* (London, 1983), 205–30. This writer is also the author of authoritative work on the early history of the British Association.

15. For a popular interpretation, see R. Little and W. Reed, *The Confucian Renaissance* (Sydney, 1989).

16. D.N. McCloskey, *If You're So Smart: The Narrative of Economic Expertise* (Chicago, 1990).

17. A. Koestler, *The Act of Creation* (London, 1964).

18. Schaffer, *op. cit.* (1).

19. T.S. Kuhn, *The Structure of Scientific Revolutions* (Chicago, 1970).

20. I. Inkster, 'Intellectual Dependency and the Sources of Invention: Britain and the Australian Technological System in the Nineteenth Century', *History of Technology*, 1990, 12: 40–64.

21. Various distinctions have been made between incremental and breakthrough changes in the history of technology, these not being equivalent to the older dichotomy of invention and innovation. Thus Joel Mokyr's microinventions are 'improvements' resulting from 'search and inventive effort, and respond to prices and incentives ... Macroinventions, on the other hand, do not seem to obey obvious laws, do not necessarily respond to incentives and defy most attempts to relate them to exogenous economic variables. Many of them resulted from strokes of genius, luck or fortunate misunderstandings': Mokyr, *Twenty Five Centuries of Technological Change: An Historical Survey* (London, 1990), 7. However, although for Mokyr, Watt's work was 'paradigmatic', comparable to that of Newton in physics and precursor to 'an economic revolution', elsewhere and, I think, uneasily, Mokyr defines the entire development of the steam engine from Papin to Watt as a 'macroinvention': e.g. Mokyr, in Mokyr (ed.), *The British Industrial Revolution* (Boulder, CO, 1993), 20. A taxonomy which fails to distinguish systematically between the work of, say, Desaguliers and Watt may not be of great utility to empirical historians of technology and at present most analytical distinctions seem to operate on too many, very mixed criteria at once, ranging from the notion of cognitive dissonance to that of economic impact.

22. D.S.L. Cardwell, *Technology, Science and History* (London, 1972), especially Chapter 4, and his *Fontana History of Technology* (London, 1994), Chapters 5, 7 and 19.

23. For a quite brilliant and entertaining introduction, see Preston King, 'Historical Contextualism: The New Historicism?', *History of European Ideas*, 1995, 21: 209–33. The 'context' is not magesterially privileged.

24. J.P. Muirhead, *The Life of James Watt* (London, 1858): quoting Watt's relay to John Hart, p. 82.

25. D. Lamb and S.M. Easton, *Multiple Discovery* (London, 1984), 16–19.

26. D. Fleming, 'Latent Heat and the Invention of the Watt Engine', *Isis*, 1952, 43: 3–5.

27. R.E. Schofield, *The Lunar Society of Birmingham* (Oxford, 1963), 62.

28. Muirhead, *op. cit.* (24), 61.

29. Watt to Brewster, quoted in J.P. Muirhead, *The Origin and Progress of the Mechanical Inventions of James Watt, Vol. 2* (London, 1854), 355. Also, Watt to Robinson on Black's

influence in letter 16 December 1799: 'he taught me to reason and experiment in natural philosophy', in E. Robinson and D. McKie, *Partners in Science: Letters of James Watt and Joseph Black* (London, 1970), 319.

30. Watt to Hutton, December 1795 in J. Jones, H.S. Torrens and E. Robinson, 'The Correspondence between James Hutton and James Watt', *Annals of Science*, 1995, 52: 357–83, p. 380.

31. Schofield, *op. cit.* (27), 64, 66.

32. A.R. Hall, *From Galileo to Newton 1630–1720* (London, 1963), 333; J. Needham, *Clerks and Craftsmen in China and the West* (Cambridge, 1970), 136–202, e.g. 202.

33. J. Cohen, 'Can Human Irrationality Be Experimentally Demonstrated?', *Behavioural and Brain Sciences*, 1981, 4: 317–70; S. Stich and R. Nisbett, 'Justification and the Psychology of Human Reasoning', *Philosophy of Science*, 1980, 47: 188–202; S. Stich, 'Rationality', in D.N. Osherson and E.E. Smith (eds), *Thinking: An Invitation to Cognitive Science, Volume 3* (Cambridge, MA, 1990), 172–96.

34. The best material on which in the English language is now collected in the new Economic History Society collected volumes of reprints, *The Industrial Revolutions*, general editors R.A. Church and E.A. Wrigley (Blackwell, Oxford), including volumes on Britain, Europe, North America and Japan.

35. *Op. cit.* (14), 254.

36. I. Inkster, *Science and Technology in History: An Approach to Industrialisation* (London, 1991).

37. Material derived from my continuing work on international patenting prior to 1914. Prior to World War I such consolidated sites of engineering and legalistic activity served to routinize the reaction to novelty through the exhaustive presentation of appeals, the redesigning of 'crucial ideas' (as the law offices of J. Walker Douglas of Philadelphia expressed it), the investigation of 'practical merits' and the provision of special financial schemes for potential patentees.

38. 'The Coal Tar Colour Industry in Germany and England', *Nature*, 12, December 1901: 138–9; *Board of Trade Journal*, 3 February 1910: 136; *Twelfth Report of the Controller-General of Patents, Designs and Trade Marks with Appendices for the Year 1894* (London, 1895); *Ding Polytechnic Journal*, 1894, 295: 160–4.

39. *Nature*, 1901, 65: 138.

40. Reprinted in A.T. Simons, 'Education and Progress in Japan', *Nature*, 1904, 69: 416–18.

41. 'Why Japan Is Victorious', *Nature*, 1905, 72: 128–9.

42. I. Inkster, 'Technology Transfer and Industrial Transformation: An Interpretation of the Pattern of Economic Development circa 1870–1914', in R. Fox (ed.), *Technological Change* (Oxford, 1995), 177–200.

43. Technical details may be investigated through the contemporary expert engineering accounts 'Iron and Steel Works – Reschitza, Hungary', *The Engineer*, 11 January 1884: 23–26, 18 January 1884: 58–9, 1 February 1884: 85–6, 15 February 1884: 134–5, 29 February 1884: 159–61. For background, see I.T. Berend and G. Ranki, *Hungary: A Century of Economic Development* (New York, 1974).

44. L. Katus, 'Economic Growth in Hungary during The Age of Dualism 1867–1913', in E. Pamlenyi (ed.), *Socio-economic Researches in the History of East Central Europe*, no. 62. (Budapest, 1970); W.W. Rostow, *The World Economy, History and Prospect* (London, 1978): 74.

45. For an unusual and neglected attempt, see D. Senghaas, *The European Experience: A Historical Critique of Development Theory* (Leamington Spa, 1985).

46. Links between late development, foreign technology transfers and political outcomes is still an under-researched area, opening a potentially huge opportunity to historians of technology. A key matter relates to time horizons: political control mechanisms designed to sustain a rapid technological modernization programme in the face of displacement, unrest and disturbance might succeed in the medium term (i.e. 'allow' an 'industrial revolution' to occur successfully), yet in the longer run give rise to dire political-ideological outcomes. For varied assays of this topic see B. Moore Jr, *Social Origins of Dictatorship and Democracy* (London, 1969); P. Anderson, *Lineages of the Absolute State* (London, 1979); I. Inkster, 'Conflict and Conflict Management: The Other Side of Meiji; in G. McCormack

and Y. Sugimoto (eds), *The Japanese Trajectory: Modernisation and Beyond* (Cambridge, 1988), 107–128; I. Inkster, 'Relative Backwardness and Revolution: A Note on Marx, History and the Transition to Socialism', *Journal of Contemporary Asia*, 1992, 22: 146–51; K.A. Chaudhry, 'The Myths of the Market and the Common History of Late Developers', *Politics and Society*, 1993, 21: 245–74.

47. See essays by Macpherson, Crawcour, Smith, Inkster, Rosovsky, Dore and Yamamura in W.J. Macpherson (ed.), *The Industrialisation of Japan*, in the series Economic History Society, *The Industrial Revolutions, Vol. 7* (Oxford, 1994), as per note 34 above.

48. I. Inkster, 'Culture, Actions and Institutions: On Exploring the Historical Economic Successes of England and Japan', in P. Gouk (ed.), *Wellsprings of Achievement: Cultural and Economic Dynamics in Early Modern England and Japan* (Aldershot, 1995), 239–66.

49. For which see papers by Totten and others in the special issue of *The Developing Economies*, 1977, 15; I. Inkster, *Science, Technology and the Late Development Effect; Transfer Mechanisms in Japan's Industrialisation circa 1850–1912* (Tokyo, 1981).

50. N. Umetani, *The Role of Foreign Employees in the Meiji Era in Japan* (Tokyo, 1971).

51. Inkster, *op. cit.* (49), Chapter 4.

52. See Chapters 5 and 6 of H.J. Jones, *Live Machines, Hired Foreigners and Meiji Japan* (Vancouver, 1980).

53. See especially pp. 171–80 of I. Inkster, 'On Modelling Japan for the Third World', *East Asia*, 1983, 1: 155–87.

54. M.M. Pitt and L.F. Lee, 'The Measurement and Sources of Technical Efficiency in the Indonesian Weaving Industry', *Journal of Development Economics*, 1981, 9: 43–64; M. Nishimuzu and J.M. Page, 'Total Factor Productivity Growth, Technological Progress and Technical Efficiency Change: Dimensions of Productivity Change in Yugoslavia 1965–78', *The Economic Journal*, 1982, 92: 920–36.

55. D. Landes in Mokyr, *op. cit.* (7), 166.

James Watt, Mechanical Engineer

R. L. HILLS

We find in Matthew Boulton's first letter to James Watt, written on 7 February 1769, the vision of erecting a steam engine manufactory where they would produce steam engines for the whole world. This part of the letter is reproduced here.

> I presumed that your engine would require money, very accurate workmanship, and extensive correspondence, to make it turn out to the best advantage; and that the best means of keeping up the reputation, and doing the invention justice, would be to keep the executive part out of the hands of the multitude of empirical engineers, who, from ignorance, want of experience, would be very liable to produce bad and inaccurate workmanship; all of which deficiencies would affect the reputation of the invention. To remedy which, and to produce the most profit, my idea was to settle a manufactory near to my own, by the side of the canal, where I would erect all the conveniences necessary for the completion of engines, and from which manufactory we would serve all the world with engines of all sizes. By these means and your assistance we could engage and instruct some excellent workmen, who (with more excellent tools than would be worth any man's while to procure for one single engine) could execute the invention 20 per cent. cheaper than it would be otherwise executed, and with as great a difference of accuracy as there is between the blacksmith and the mathematical-instrument-maker. It would not be worth my while to make for three counties only; but I find it very well worth my while to make for all the world.[1]

Boulton assessed the position correctly when he realized that money, accurate workmanship and extensive correspondence would be needed to make the engine business a success. He was also correct in realizing that high engineering standards might be best attained by keeping control of production themselves and that mass production would reduce the price. Yet neither Matthew Boulton nor James Watt ever established a place where they themselves built complete engines. This partnership always remained one of essentially consulting engineers and it was left to their sons to set up the Soho Foundry in 1795. This chapter suggests that one reason for this was the experience James Watt gained in Scotland as an engineering consultant, which showed that steam engines could be built without having an extensive engineering works. We will examine Watt's career in Scotland first as mathematical instrument maker and then engineering consultant, concentrating on what today we would call the mechanical engineering aspects of these businesses.

As a lad, Watt must have had some interest in or aptitude for mechanical pursuits. Without any archival records to confirm the anecdotes recorded in Watt's earlier biographies, it is very difficult to determine what is fact. That he had built some models as a child is confirmed through the discovery of some in the attic of his father's house in 1801. By that time only a model of a house and 'a crane or something of that nature' survived,[2] which is far fewer than the miniature pulleys or blocks, pumps, capstans, a crane and a barrel organ recorded by Williamson.[3] Watt was given some carpenter's tools and his own forge so that he helped in his father's shipwrighting business, taking particular interest in the various nautical instruments.

Watt must also have inherited something of the abilities of his grandfather, who had been a teacher of mathematics at Cartsdyke. However, this interest was not aroused until he was 13 or 14. At first he appeared dull at school until he was put in the mathematical class, and from that time he made rapid progress.[4] He showed great aptitude in the study of geometry.[5] Two rough notebooks have survived, which, from the scrawly, immature writing, were probably copied by the young Watt from some other book. Their interest for us lies in their contents, which show the lines along which he was thinking. The first contains notes on 'Surveying of land',[6] with instructions on instruments, how to measure land, find distances and levels etc. The second is on 'Dialing and making instruments'.[7] Here we find drawings of various instruments including scales, dialing tables and 'How to make a sea quadrant – This instrument is ye most easiest divided instrument of any for it consists all of equal parts.' We can well understand how the young Watt would have been fascinated with scales and the latest mathematical instruments, which would have been to him the equivalent of the computer in our present day, and the quadrant held a special place in his interests.

We know that Watt was sent to his mother's relatives in Glasgow for a year in 1754 when he was 18. Tradition has it that this was in order to learn the trade of a mathematical instrument maker and that only an optician could be found in Glasgow to teach him. This person 'was a sort of Jack-of-all-trades, who sold and mended spectacles, repaired fiddles, tuned spinets, made and repaired the simpler instruments used in mechanical drawing, and eked out a slender living by making and selling fishing-rods and fishing tackle.'[8] If indeed Watt did receive any training in Glasgow, the name of his first master remains unknown. Bryden has pointed out that there were at least two men to whom he could have been indentured: Henry Drew and George Jarden, who acted as laboratory stewards to successive professors of natural philosophy in the university and were members of the Incorporation of Hammermen, with special competence in horological work.[9] Later Watt had some financial dealings with George Jarden when he was moving to Glasgow at the end of July 1757, so they must have known each other.[10]

We must ask why Watt might have been sent to Glasgow. His recently widowed father would not have found it easy to look after two teenage lads, especially as at some time around this period he suffered the loss at sea of a ship. The most natural career for James Watt to have followed would have been that of his father, and we can understand that it would have been beneficial for all concerned to have a son trained as a merchant at his relatives in the town where much of his father's trade originated. The first letter Watt sent back to his father from Glasgow on 11 June 1754 contains an invoice of goods bought there for his father; tanned skins, chest lox [*sic*], padlocks, hooks, marble paper, pencils etc.[11] The other five surviving letters which Watt sent to his father between then and 9 May 1755 all contain details of purchases for his father or payments from other people, with nothing about making mathematical instruments. Presumably, part or most of his time in Glasgow was spent acting as a sort of agent for his father.

That he went to Glasgow to be something more than just an apprentice is also suggested by the list of items he took with him. The almost new shoes, the ribbed stockings and plain ones in grey blue, blue and white or silk, 'rufled shirts', '8 Holland night caps', black silk cravat, one black and one blue coat, two black cloth, one cut velvet and one tartan waistcoats, are all clothes which someone representing his merchant father might wear. But in addition we must note the 'lether apron' and tools such as a jack, a hand 'plain', a saw, 'chissals, gamlets, fyles' and '1 Quadrant & Vanes'.[12] The quadrant has appeared again.

This dual pattern of the young Watt acting as his father's agent while at the same time learning the business of mathematical instrument manufacture we can discover while he was in London. Watt arrived there on 18 June 1755.[13] On 21 July, before he had made the final arrangements to work for John Morgan, he wrote home,

> You can have scales here at two prices to wit at 12/- pr. Doz & 15/- pr Doz so will send some of Both. I find we have been overcharged by Greg & Campbell for most of the things in ye way ... You'll please send up an acct. of what things more you want here which will see to buy as cheap as possible ... This is very bad write my hand shaking after working.[14]

Greg & Campbell were wholesale merchants with whom Watt senior had been dealing and Watt tried to find cheaper sources of supply to by-pass them. He arranged shipment to Glasgow and generally acted as his father's London agent while he was there. This sort of activity he continued after he had returned north and set himself up as mathematical instrument maker in the University of Glasgow. There is no indication to show whether the young Watt received any financial benefit from these activities, but it is possible that he hoped to repay to his father the money spent on him for his London training through these merchanting ventures, and they may have helped to sustain him during the initial months in Glasgow.

We have a list of the equipment which Watt intended to purchase in London and take back with him to Glasgow. It consisted of virtually all hand tools and included a standing vice, anvils, hammers, files, screw plates, dividers, a large turn bench, an alphabet of letters as well as materials for manufacturing instruments.[15] He also intended to bring down the wooden parts of three different types of Hadley's quadrants which he would copy because he had 'never wrought at them' and he was going to make a dividing plate for producing them.[16] In addition, he returned home with 300 copies of a booklet about Hadley's quadrant which shows the sort of instrument he hoped to sell.[17] The backstaff was transformed into the octant by John Hadley, who published his invention in the *Philosophical Transactions of the Royal Society* in 1731. The novelty was the use of a mirror mounted over the pivot of a radial arm that moved over a graduated arc. Viewing was through a pin hole sight at a half-silvered mirror, while the horizon could be seen through the clear half, so 'bringing the sun down' to the horizon. The name of this instrument causes confusion because it can measure 90 degrees, a quadrant of a circle, but the arc through which the arm moves occupies only one-eighth of a circle, an octant.[18] This instrument came into general use after 1750, and presumably Watt had identified the commercial prospects for selling it to the increasing number of ships sailing in and out of the Clyde. At this period, such instruments would have been made from wood.

Watt wrote to his father from London on 19 August 1756, just before he set out for Greenock and home, about borrowing more money, probably for stock items which included telescopes.[19] During that autumn he was in Glasgow and repaired the observatory instruments for the university which had been sent from Jamaica.[20] Then he set himself up in business in Greenock and his 'Waste Book' has survived. An inventory for 3 January 1757 contains items to the value of £21.4.2 and includes a complete Hadley's quadrant.[21] That month and into February, Watt was busy with buying parts for a 'great lathe', which is the first indication we have of any machine tool among his capital assets. Most of the other entries refer to making small parts for navigation instruments such as Hadley's quadrants and also to buying more hand tools from John Wylne. While at Greenock, Watt had sufficient work to employ Alexander Mackenzie for 23½ days during May and June 1757, apparently on a casual basis.[22]

Just under a year from leaving London, on 2 August 1757, there is an entry in the Waste Book, 'Expenses removing to Glasgow, Boat freight 5/-, porterage 1/6, cartage, 1/2, town dues, 6 d., Brushes 1/- & pr Buckles 6/-, Pocket Expenses 1/-, Physic & drawing a tooth 3/6.'[23] Here Watt would expand from a struggling one-man tiny workshop into a considerable business. Already he may have had his eyes set on wider horizons because one of the things he took with him from Greenock and which he appears to have erected there was his 'great lathe'. Through his father, he could retain his trading connections at Greenock while Glasgow presented him with the opportunity to diversify through the university connections. In fact, he pointed out to his father in October 1757 that

> There is no occasion to send to London whether for Davis or Hadleys Quadrants as hope to have some of both soon. If you can get a good well shaped Davis quadrant that has been approved by some Experimental seaman send it up & shall return it in a week, at any rate send the best you can get. If the Compass boxes be ready send them up & shall see to get some of them done.[24]

At this stage in his career, Watt was willing to copy instruments manufactured by others to build up his stock-in-trade.

Watt was enquiring about 'any lad that can file tolerably well that would be willing to come here' in February 1758,[25] and the accounts show that some people were employed on a casual basis during that first year. That May, he was very busy but progress was slow and he took stock of his position in the following September:

> As I have now had a year's trial here I am able to form a judgement of what may be made of this Business & find that unless it be the Hadleys Insts there is Little to be got by it as at most other Jobs I am obliged to do the most of them my self & as it is impossible for one person to be Expert at everything they very often cost me more time than they should do However if there could be a ready sale procured for H. Quadrants I could do very well as I & one Lad can finish 3 in a week easily selling them at 28/6 d which is vastly below what they were ever sold at before I have 40/ clear on the three so it will be absolutely necessary that I take a trip to Leverpool to look for Customers & hope that upon the profits of what I shall be able to sell there I can go to London in the spring where I make no Doubt of selling more than I can get made all which I want your advice on, and if that does not succeed I must fall into some other way of Business as this will not do in its present situation.[26]

Watt was finding himself in the classic dilemma of the small entrepreneur. It was easier to do the work himself than to try to explain and supervise someone else, but this did not pay. Specialization and batch production were more profitable, but then sales outlets had to be found for any production.

Watt's letters to his father at this period show also that working capital was short. This was probably not helped by a venture into publishing a map of the Clyde, which his uncle had surveyed some years before. More profitable was a partnership with Alexander Wilson and Joseph Black in an astronomical clock, out of which Watt received £18 in May 1759.[27] How successful a trip to London that August was we do not know.[28] We have accounts for the purchase of stock to sell in Glasgow but no records of any of his production sold in London. The list of his purchases reads as follows:

Chris. Steadman, as per Invoice	19.15. 0
Bought sundrie Optical do. from J. Tomlinson,	17.18. 3
do. from Lincoln as per Invoice,	6. 7. 0
Sundrie rules from R. Rust, as per do.	14. 1. 6
Sundrie Sea Compases from J. Jarman,	4. 8. 6
Globes from N. Hill, [globemaker]	3. 2. 6
? fiskin Cases from R. Bounds,	2.16. 0
Spectacles from J. Oakley,	1. 7. 0[29]

Tomlinson and Lincoln were opticians and Bounds was a 'case maker', probably for spectacles. That September, back in Glasgow, Watt was selling three microscopes, four each of Davis and Hadley's quadrants, three dozen Gunter's scales, twelve sea compasses and two hanging compasses. Business must have been flourishing.

The real change in Watt's business fortunes began in October 1759 when he formed a partnership with an architect, John Craig. Watt by himself had not done too badly, for the stock had increased to £108.0.8½[30] and there was also £91.19.3½ in credits by various sundries. However, the figures we have for Watt's accounts are so incomplete that it is impossible to draw any firm conclusions. Craig was willing to lend Watt, or the business, small sums of money on interest. Between 15 May 1760 and 13 November 1764, this seems to have totalled £700. This enabled Watt to move out of the restricted premises within the university, for the partners set up a shop in the Saltmarket. Watt received a salary of £35 per annum and shared the profits.[31] From this period, the whole nature of Watt's business changed.

Business at this shop flourished and branched out into new lines, including that of manufacturing musical instruments. Watt did not know one musical note from another, but mastered the theory of music so that he could proportion his instruments correctly. The earliest mention of anything to do with this trade that has been discovered so far has been the payment by John Campbell, a merchant in Glasgow, on 13 December 1760 for '4 Screws to Double Bass for Musical Society, £1.1.0'.[32] Later, Campbell paid 8/- to have two flutes mended on 30 November 1761 and in January 1763 bought a flute for 10/-. By that time, manufacturing flutes had become a profitable line of business which Watt continued for some years. In December 1761, he corresponded with Alexander Cumming in London about buying sets of pipes for an organ. Organs began another production line and he received £20 in June 1763 as part payment for one from the Royal Arch Masons Lodge in Glasgow.[33] In March 1764, he received a further £24 'in ready money for an organ from Mr Lockhart',[34] and he made a portable organ for Joseph Black in 1768.[35] One of his organs survived at St Andrew's Parish Church when Dickinson was writing his biography of Watt in 1935.[36] Other types of instruments which Watt made were guitars (with this trade starting in the summer of 1762), mandolins and violins at about the same time.

Most of these musical instruments and the mathematical instruments were made from wood and therefore could have shared similar production techniques and could have employed the same people to manufacture them. Two 'Journey Mens Books' survive, covering payments to Watt's workforce from the period 6 October 1760 to June 1771, but, once again, these records are very fragmentary. The accounts open with five people being paid, of whom one ceases in March 1761, but another, John Gardner, continues until the end of 1770 and will become very important later. During 1761, the ranks of journeymen were swelled by a further six, with another four in 1762, three more in 1763 and three in 1764. There may have been an initial two-year contract but, by 1764, the records are too incomplete to derive a clear picture. Dickinson says that there were 'sixteen at the finish' and Watt had even taken apprentices.[37] That the work-force built up to these sort of numbers is confirmed by an entry, 'Paid Journeymens wages preceding the first of January 1764' which shows that a total of £201.2.11½ was paid out to 16 people, presumably for the past year's work.[38] Wages ranged from 5/- to 3/- per week, but some payments may have been made for contract work making specific parts. Of those listed in the 'Journey Mens Books' who have some trade associated with their wages, four were involved with musical instruments, one with scientific instruments, one with both scientific and musical instruments and one with glass grinding. John Gardner made musical instruments but later became involved with steam engines. He took over Watt's instrument business, which he and his family continued to run for many years.[39] In addition there was James Syms, who may have been one of Watt's shopkeepers.

By this time, Watt must have diverted most of his attention to the managerial side of the business rather than the actual production. Some evidence for the methods he employed can be determined from surviving artefacts in his Garret Workshop, now preserved at the Science Museum in London. In 1765 he developed his perspective drawing machine. From boxes and other incomplete parts for this apparatus in his workshop, he must have produced these machines in batches, which would have helped to speed production. Then small devices to assist in making different parts of other production lines can be found as well. There are various gauges which must have helped in the making of standardized parts. For cutting the swells inside flutes beneath the finger holes, there is a special rotary file which can make three at the same time. For the lines on scales, there is a scriber which can draw ten at once. There is also a sector which has been engraved but the two arms are not separated and the hinge was never finished, which could have acted as a template or master for making others. Plates for printing barometer scales on paper exist, a much cheaper method than engraved brass plates.[40] Just how far such labour-saving devices can be attributed to Watt himself is now impossible to determine, but it is probable that many of these were his own ideas because in later life he was continually trying to improve his instruments, steam engines etc. It seems safe to assume that Watt introduced many methods of improving his production techniques in his scientific and musical instrument business, which helped to give him a lead over his competitors. This was a period when brass was replacing wood in many instruments. Lying in a box are cast brass frames for making a couple of surveyor's quadrants which were never finished. These raise the question of how Watt, living in remote Glasgow, was able to keep abreast of the latest designs and advances in instruments which would have originated in London. He could quickly have fallen behind in such a competitive race because instruments were developing quickly at this period.

The room in the university precincts was quite inadequate for this expansion in Watt's work-force and his account books show that he occupied some other building as a workshop. Unfortunately we do not know where it was, what benches or tools it might have contained or for how long he rented it. In the autumn of 1763, he branched out into another venture, that of a general shop on quite a large scale. That June he borrowed £150 from Joseph Black[41] and, in the following August, £42 and £112.5.0[42] from John Craig. With this money, he went on a purchasing spree to London, returning through Sheffield and Birmingham. There was no contact with Matthew Boulton at this period. Watt laid out at least £200 on items such as snuff boxes, corkscrews, spectacles, nut crackers, china bowls, red ribbon, kitchen towels, keys, broaches, theodolites, pair pistols, gun, compasses, measuring tapes, shoe buckles, knee buckles, sleeve buttons, dice boxes, knives, forks and much more. An advertisement appeared in the *Glasgow Journal* of 1 December 1763. 'James Watt has removed his Shop from the Sautmercat to Mr Buchanan's land on the Trongate where he sells all sorts of mathematical and musical instruments, with a variety of Toys and other goods.'[43] This 'toy' shop was quite an extensive operation, so that, when he was only 28, Watt had a string of business successes to his credit. He had launched his mathematical instrument business and followed that by his musical instrument business; he had a workshop employing around 16 people, of whom some were probably on short-term contracts, in a manufacturing capacity, and he had this large general merchant's shop. At this period of his life, he called himself 'merchant'.

Within the next three or four years, Watt had altered the direction of his career completely. It was during the winter of 1763 to 1764 that he was asked to repair the model of a Newcomen engine which belonged to the Natural Philosophy Class of Glasgow University. This was to lead to his inspiration, in the spring of 1765, of the separate condenser, on which his subsequent fame was to be founded. If his new engine were to be a practical success, Watt must have quickly realized that he needed to gain experience of designing, erecting and running ordinary atmospheric steam engines. He launched out into this field in the autumn of 1765 and this story will form the conclusion of this paper. But either just before or on 4 December 1765, John Craig died.[44] While John Roebuck would take over responsibility for Craig's loans to Watt, it has been suggested that Craig looked after the financial side of Watt's businesses and, without his guiding hand, Watt's little empire seems to have started to collapse.

Signs of trouble appear soon after Craig's death, because Watt wrote to his father on 23 January, 1766.

> I am much obliged to you for the trouble you have been at in procuring pate for me I will readily give him the £4 – & as I shall be obliged to go out of town on Wednesday next & shall stay a few days therefore wish that pate could be here on Monday tho I have no stronger proof than suspicions against James yett I think it proper to discharge him which I have accordingly done in a civil manner he is to stay here 2 months in which time pate will have made some proficiency Peggy is now in the shop along with him & proposes attending constantly & as I intend droping most of the Lads I shall be there a good [hole] myself.[45]

It seems that James, probably James Syms, had failed to superintend the shop properly and was to be dismissed. Watt's wife, Margaret, had been called in to help out and Pate was to be trained to act as both shopkeeper and accountant. But the more significant comment is the one about 'droping' most of the lads. This term 'lads' appears to have been used by Watt about his journeymen, so that it seems that he was already contemplating running down his workshop. Perhaps his production of musical instruments had already glutted what must have been quite a small market. We do not have enough evidence to plot the gradual contraction of this side of his activities, so we do not know when the workshop was finally closed.

His 'toy' shop too seems to have continued to suffer from lack of supervision. In 1765, Watt became involved with Black and Roebuck in trying to produce alkali from sea salt. He may have been acting as technical adviser to the Delftfield Pottery in Glasgow from the beginning of 1763,[46] but February 1768 is a much more likely date for this, when he told Black that he had purchased part of a share in this concern.[47] But it was the beginning of Watt's career as a civil engineer that really caused this change of direction. Jenkins says that in the summer of 1766 Watt gave up his shop (which is definitely incorrect) and took an office in King Street where he set up as a land surveyor and engineer.[48] No reference is given but this was the period when a canal was being proposed to link the Clyde and the Forth, and it was around this time that Watt began to call himself 'Engineer'. Watt may have worked under Robert Mackell in October 1766 on a survey of a northerly line following the River Forth itself nearly to Aberfoyle, before the canal would have turned south-westerly and crossed the high ground before dropping down to Loch Lomond.[49] He certainly helped to survey a southerly route during February 1767, which would have passed through the coal fields east of Glasgow and entered the Clyde to the south-east of that town. He was offered £100 to go to London to support the Bill for this 'ditch', as its opponents called it, during that spring. Mrs Watt had to move house during May and from this we learn that Watt had some sort of garret workshop for his own experiments.[50] The withdrawal of this Bill from Parliament opened the way for Watt to survey the line of what became a little later the Monkland Canal through these same coal fields.

These absences on surveys caused his father to send him a letter warning him about the state of the shop.

> Its unserten how fare yt small canall will give content to ye London merchts. I think you should keep yourself as mutch disingeged as you cane from atendence on ye executing of it as your helth may be impaired with such weather as we have . . . tho you get £100 for going to London & atending their some time it may be a furder loss to your shop or town bussiness.[51]

His father also pointed out the need to secure proper agreements concerning terms of payment for this survey work, something which his son failed to do in many of his surveying contracts.

The prospects of life as an 'engineer', what we today would call a civil engineer, must have seemed attractive to Watt through both the enhanced financial prospects it held out and a more congenial lifestyle. Life in the open air seems to have improved his health and his appointment as part-time resident engineer on the Monkland Canal in the summer of 1770 with a salary of £200 per annum must have seemed a very advantageous proposition. During the spring of 1770, Watt had been carrying out trials on his own engine at Kinneil and had probably just begun to make his separate condenser work properly. He left there on 8 April to begin a survey of the Strathmore Canal near Perth, never to finish those trials on his engine at Kinneil because Roebuck was declared bankrupt in that June.

Watt had had to decide whether to try to continue with his own engine or take up the career of civil engineer. We can well understand his dilemma, which he expressed to his friend William Small that September 1770.

> I had just got to Kinneil to finish my experiments on the engine, when, on a day's warning, I was called to Glasgow to attend a survey of the river [Clyde with John Golborne]. Before that was finished, the canal [Monkland] I had projected last winter, for which an Act had been obtained, was wanted to be begun under my inspection. I had now a choice whether to go on with the experiments on the engine, the event of which was uncertain, or to embrace an honourable and perhaps profitable employment, attended with less risk of want of success: – to carry into execution a canal projected by myself with much trouble, or to leave it to some other person that might not have entered into my views, and might have had an interest to expose my errors; (for everybody commits them in those cases.)
>
> Many people here had conceived a much higher idea of my abilities than they merit; – they had resolved to encourage a man that lived among them rather than a stranger. If I refused this offer I had little reason to expect such a concurrence of favourable circumstances soon. Besides, I have a wife and children, and saw myself growing gray without having any settled way of providing for them.[52]

Here we see Watt's reasons for accepting the post on the Monkland Canal: the financial prospects, the need to support his family and the uncertainty of the success of his own steam engine. Three years later, the positions were reversed. The civil engineering career was falling into tatters with the collapse of the Ayr Bank in 1772 and the slump that followed, the termination of the Monkland Canal contract with part of his salary unpaid and the dim prospects for other work, such as the Crinan or Caledonian Canals. Boulton's offer of financial support for Watt's own steam engine must have shone like a ray of hope in a depressing situation compounded by his wife Margaret's death on 24 September 1773. When Watt arrived in Birmingham in May 1774, he could not have known that he was destined to live there for the rest of his life.

It was in Watt's employment on the Monkland Canal that we can see him changing from being a direct supervisor of engineering work into a consulting engineer where he could draw up the initial plans and leave the execution to others. He would also apply this to his mechanical engineering interests on the steam engine. Watt's problems with running his shop have been discussed earlier, but we have no hints about similar problems with his lads or journeymen. Yet labour relations loom large on the Monkland Canal and were one of the reasons for his becoming a consulting engineer.

Watt had realized that he was out of his depth in supervising the actual construction of the Monkland Canal when he wrote to Small in September 1770.

> Nothing is more contrary to my disposition than bustling and bargaining with mankind: – yet that is the life I now constantly lead. Use and exertion render it rather more tolerable than it was at first, but it is still disagreeable. I am also in a constant fear that my want of experience may betray me into some scrape, or that I shall be imposed upon by the workmen, both which I take all the care my nature allows to prevent. I have been tolerably lucky yet.[53]

Two years later, the situation had not improved despite the experience which Watt ought to have gained during the interval. He poured out his soul to Small again, but this time we can see how he proposed to work in the future.

> I decline only being the manager, and not being engineer. I wrote you before how grevious that first part of the business was to me, and it daily becomes more so. Everything has been turned over upon me, and the necessary clerks grudged to me; I am also indolent, and fearfully terrified to make bargains, and hate to settle accounts. Why therefore shall I continue a slave to a hateful employment, while I can otherwise, by surveys and consultations, make nearly as much money with half the labour, and, I really think, with double the

> credit? for a man is always disgraced by taking upon him an employment he is unfit for. I have no quality proper for this employment but honesty, which reproaches me for keeping it so long.
>
> Remember in recommending me to business, that what I can promise to perform is, to make an accurate survey and a faithful report of anything in the engineer way; to direct the course of canals; to lay out the ground, and to measure the cube yards cut, or to be cut; to assist in bargaining for the price of work, to direct how it ought to be executed, and to give my opinion of the execution to the managers from time to time. But I can upon no account have anything to do with workmen, cash, or workmen's accounts, nor would I choose to be so bound up to one object that I could not occasionally serve such friends as might employ me for small matters.[54]

Another letter in a similar vein followed a few days later in which we can clearly see Watt's fear of dealing with workmen.

> I am extremely indolent, cannot force workmen to do their duty, have been cheated by undertakers and clerks, and am unlucky enough to know it. The work is done slovenly, our workmen are bad, and I am not sufficiently strict . . . What provokes me most is, that I am sensible that most people could in the same time have done much more and better work, possibly with as little trouble to themselves. I would rather face a loaded cannon than settle an account or make a bargain. In short I find myself out of my sphere when I have anything to do with mankind; it is enough for an engineer to force Nature, and to bear the vexation of her getting the better of him. Give me a survey to make, and I think you will have credit of me; I can draw tolerably; set me to contrive a machine, and I will exert myself.[55]

If this point about Watt's fear of his work-force has been emphasized too much, this has been done because it may help to explain why Boulton and Watt never established a complete manufactory for steam engines in Birmingham, but left this to their sons. Watt's way of escaping from this dilemma was to turn himself into a consulting engineer. He would assess the situation, he would draw up the plans for carrying out the scheme, but he would recommend a contractor to do the work. In this way, as he pointed out to Small, he could still gain the credit for the scheme but avoid the hassle.

Two of Watt's schemes show us how he intended to operate as a consulting engineer. The first was for a new bridge over the River Clyde at Hamilton. Some trial borings were made during the summer of 1770 to ascertain the nature of the river bed.[56] In the ensuing months, Watt must have drawn up his plans, which included 'Regulations for the Undertaker of Hamilton Bridge'.[57] Unfortunately these are undated, but 20 paragraphs give directions for the piling for the piers, the construction of the arches and so on. Estimates for the stonework and hewing amounted to

£1202.3.0 and for the piling and foundations to £950.4.2. Tenders must have been sought and the Duke of Hamilton's tutors (the Duke was a minor at this time) were going to ask Watt to supervise the actual construction, but he refused and recommended a mason from Perthshire.[58]

In the second scheme, Watt stated his method of working to the Lord Provost of Glasgow when asked to carry out repairs to the dry dock at Port Glasgow in August 1771.

> I would undertake to direct the repairs in the following manner. I would make agreements with different contractors for the several branches of the repairs, & would apoint some proper person to overse the execution masonry or any other part the faults of which could not be seen when finished. The other works I would review myself when finished & see that they were properly executed in terms of their respective agreements. The town of Glasgow are to be bound to advance the money to carry on the works as required & I to deliver sufficient vouchers for these disbursements for which service I trust the good Town will allow me a reasonable gratification.[59]

Watt would draw up the initial plans, and contractors would be appointed with a proper supervisor, so that Watt himself remained remote from direct contact with the work being carried out. Here was his ideal role as consulting engineer, but he failed to state his own terms of employment.

We can trace the development of Watt as a consulting engineer in the mechanical engineering aspects of his work as well through his construction of Newcomen steam engines. Before he left Glasgow for Birmingham in 1774, he had erected a number of such engines without needing to have any capital investment in workshop facilities whatsoever. He was able to do this partly because the Carron Company had been founded in 1759 by two Englishmen, Samuel Garbett and John Roebuck, aided by the Scot, William Cadell.[60] For the first time in its history, Scotland had a heavy engineering works and its foundries supplied parts for steam engines. Watt was able to take advantage of its facilities both for his own engine and for the Newcomen engines with which he was involved.

The Carron Company was involved in the steam engine business at least by the beginning of 1765 when, at the end of that January, it was 'Resolved when the Loam Moulders are not employed with Goods wanted for Nu Britain or Newcastle to make 2 cylinders & 20 fathoms of pipes of the most suitable sizes to be sent to Cornwall for sale which appears to be the most probable method of introducing us into that trade.'[61] The term 'cylinder' was generally used for the steam cylinder of an atmospheric engine, while 'working barrel' meant the bored barrel of a pump. Here it is not clear whether 'that trade' refers to the steam engine business in general or to the Cornish section of it. During the later part of the summer in 1766, the company cast a cylinder possibly for Ostend, 46 inches in diameter by 10 feet long, together with the bottom and piston.[62] However, the company never seems to have built any equipment for heavy forgings because, in 1770, they had to write a letter apologizing to a colliery owner, Charles MacDowall, that they could not supply cranks for a pump operated by a waterwheel which he was installing at Machrihanish on the peninsula of Kintyre.[63]

The part the Carron Company played in the supply of Newcomen engines is revealed in a letter to Peter Buchanan, a Glasgow brewer.

> We undertake the mounting of fire Engines so far as to make the Cylinders pipes Boyler Plates etc to draw[in]g & Dimentions given us, but we never attempt to put the pipes down or do any other kind of work about the Engine further than furnishing the apparatus – For that purpose you must apply to an Engineer and in case you are not acquainted with a person sufficient for the undertaking we believe we can Recommend a Gentleman of Merit & Ingenuity to you. If you favour us with the order you may depend we shall despatch it in as short time & upon as easy terms as any other founder.[64]

That 'Gentleman of Merit & Ingenuity' was Joseph Hately, who became one of Watt's bitterest rivals in the engine business. He, and other engineers like Peter and Adam Smith, Robert Mackell and John Smeaton, could all send drawings of any parts they wanted to Carron where these parts would be cast and machined as appropriate. Watt also availed himself of this service.

It must have become clear very quickly to Watt while he was experimenting with the model engine of his separate condenser during the summer of 1765 that, if he were to construct large engines on this principle, he needed experience of designing, erecting, running and maintaining full-size engines. We have seen that, up to this date, all his expertise lay in the realm of small instruments. Smiles claimed that

> In order to master thoroughly the details of the ordinary Newcomen engine, and to ascertain the extent of its capabilities as well as of its imperfections, Watt undertook the erection of several engines of this construction; and during his residence at Kinneil took charge of the Schoolyard engine near Boroughstoness, in order that he might thereby acquire a full practical knowledge of its working.[65]

Watt's own notebooks show that it was not until 12 June 1770 that he went down the School Yard Pit,[66] but he must have looked at other engines during 1765 for, on a letter from Roebuck dated 2 October 1765, there is written, in a different hand, 'Calculation of Experiment on Old Engine' with a set of figures.[67] This addition has no date and there is a much later further addition in Watt's own hand: 'This relates to an experiment on one of Dr R.'s common engines, and served to show the great waste of fuel and steam.'

Roebuck's engines gave Watt experience in maintaining such machines and it would seem that, as Roebuck became more and more financially embarrassed, Watt tried to help him by repairing his colliery engines. Indeed, this may be one reason why Watt took so long to make his own engine run properly because his attention was diverted elsewhere. The Carron Company was advised in June 1770 that Roebuck had 'Assigned over his Effects and Management of his affairs to Messrs. Mansfield Hunter & Company and ... Thos Roebuck as Trustees for the equal benefite of his Creditors.'[68] But both before and after this date Watt was looking at Roebuck's engines. John Roebuck junior consulted Watt about problems they were having with the pumps at the West engine in November 1769.[69] This may be the same as the Wester engine, where Watt sent his 'lads' to effect some repair the following April.[70] During the summer of 1770, Watt designed a new type of beam for this engine, made out of four baulks of timber braced together in an open framing, possibly based on something he had tried at Kinneil.[71] Roebuck junior wanted Watt to look at the Dunmore engine in order to keep it running for a further couple of years while they extracted the coal reserves. During September 1770, Watt carried out tests on this engine.[72]

Watt must have acquired a good reputation for his engine work because, in April 1772, he received a request from Messrs Orrs & Wilson, Barrowfield, to inspect an engine which Peter Smith had built for them before they took it off his hands.[73] This reputation must have been based not only on his consultancy work but also on those engines which he had actually designed and erected.

Watt's 'Cash Memorandum Book' shows that he incurred expenses at Carron, Leadhills and Kennett in the autumn of 1765 and through the summer of 1766.[74] Some definitely were related to his steam engine business, which he started in partnership with Robert Mackell. Mackell was a well known surveyor and civil engineer. In 1762, he and James Murray were employed to survey a line of a canal from the River Carron across to the Clyde about five miles below Glasgow.[75] Smeaton was asked to make another survey in 1763 and there was much debate about the line and the expense. It could well be that Mackell introduced Watt to this profession when they surveyed an alternative route through the River Forth and Loch Lomond.[76] Mackell was appointed 'sub-engineer' to Smeaton for building the Forth and Clyde Canal on 14 March 1768,[77] and this took up so much of his time that he virtually gave up his steam engine business. In the autumn of 1768, he was supplied with a cylinder from Carron for an engine for Lord Perth,[78] but the Carron Letter Books, while containing a great deal of correspondence with him about canal and other matters, include nothing about any more steam engines.

The Carron Company had employed Mackell to build a reservoir 'between the Paper Mill and Dunnipace' in August 1765, which they expected would be completed by the following May.[79] They must have considered installing a Newcomen engine at about the same time to supplement their water supply in times of scarcity. 'We will in short time sett abt erecting a small engine wch by means of our waste ?Parr? wood will throw up the Water we have occasion for ... We are willing to trust the whole execution to R. Mckell.'[80] This would be the first engine built by the partnership of Mackell and Watt to a novel design.

The 'Plan & Estimate of fixing up a 72 inch Cylinder for drawing backwater at Carron . . . December 5, 1765' has survived.[81] There were to be two boilers and the water pump was 52 inches in diameter. Carron would supply the cast iron parts and the boiler plates but Watt and Mackell had to find the other parts, build the engine house and the 'canal' for the water and erect the engine. For this work they charged £998.2.6.[82] The drawing of the engine shows an 'A' frame supporting a wheel with the chain from the steam piston passing round its circumference to the pump rod on the other side. This gave a very compact type of engine which also had some form of circular steam inlet valve. However, no operating gear is shown.[83] The design is similar to a condensing engine depicted in Muirhead, with a 5-foot diameter cylinder dated to 1765,[84] and it is tempting to speculate that Watt had hoped to erect one of his own engines here, but it was not developed far enough.

This design of engine with an 'A' frame supporting a wheel instead of the conventional beam was also proposed by Smeaton in 1765, which is earlier than his Carron connections.[85] John Farey commented that 'It does not appear that this plan for a portable engine was put into execution by Mr Smeaton till some years afterwards.'[86] It would be interesting to know whether Smeaton, Mackell and Watt consulted each other over this design. It was later copied in a slightly different form by Joseph Hately in his patent of 1768.[87]

In April 1766, Watt visited the Leadhill mines for a Mr James Stirling in order to tender for an engine.[88] Drawings and complete costings have survived but Mackell was not mentioned.[89] The drawings suggest that the engine was situated down a mine shaft, so Watt proposed his compact 'A' frame type with the wheel instead of a beam. The proposal was to deepen the mine from 20 to 30 fathoms and drain it with the more powerful steam engine. When Watt visited it, the water was being cleared out by five pumps, one above the other, the highest being operated by two men at about 32 or 34 strokes a minute. From this, Watt calculated the volume of water to be removed and reckoned that a fire engine with a 44-inch diameter cylinder would be needed. The boiler would cost £19 and the cylinder and bottom £8, with the total cost estimated at £56. Another estimate was for £104,[90] but both seem low, especially when compared with the one at Carron and the next one at Kennett. From the fact that Watt's notebooks record only one visit to Leadhills, it must be doubted whether this engine was ever built.

Beginning in June 1766, Watt's cash books contain notes for 'Expenses at Kennett'.[91] Unfortunately the draft agreement between Lord Kennet and Robert Mackell, engineer at Falkirk, together with James Watt, merchant in Glasgow, is not dated.[92] They were to erect a fire engine at Pilmure with a cylinder 2 feet in diameter and 7.5 feet long, together with the boiler, pumps, engine house etc. at a cost of £400. Lord Kennet was to provide a smithy and lifting tackle. If having such a workshop provided on-site was typical of these engines, it will help to explain why engineers like Watt had no need to provide their own. This engine must have been erected because in May 1769 Lord Kennet had to write to Watt complaining about the boiler.

> I find that Mr McKell has caused make the boiler vastly too slight, so that tho the engine goes very little commonly only five or six hours in the four and twenty yet the Boiler is already wore out, so that I am obliged to make a new one, this is owing to the plates being made too thin, so that they could not last the ordinary time. I do not blame you personally for this, because I know you trusted Mr Mckell with that part of the work but as you was conjunct in the contract, it is necessary for me to complain to you of it as well as to him.[93]

This letter also shows that Lord Kennet paid Mackell for the work, so he must have been the senior partner. Watt's role in the design of the engine points to his ability in this sphere even at this early date.

Around December 1766, a letter from a Mr Dick, Glasgow, shows that Watt was involved with yet another engine, this time at Newton of Air(sic). Parts of an old engine from Kilmarnock were to be used and Watt was to replace such parts as might be necessary.[94] More important work was undertaken in 1766, when John Gardner spent 12 weeks and Murray Osborn, one of Watt's apprentices, spent three weeks at Torryburn.[95] Peter Colville, an influential colliery owner who was possibly some acquaintance of Watt's, had a coal mine there. Colville and Charles MacDowall were asked by the Carron Company at the beginning of 1767 to act as arbitrators in a dispute over how the company were mining coal in a colliery they had leased.[96] Carron asked Colville to do the same again in 1770, so his standing must have been high in the community.[97] A couple of short stroke engines were erected, showing once again Watt's willingness to try new designs. Watt undertook further work at Torryburn between August and November 1768.[98] In August 1770, he was again at Torryburn, and during that autumn Carron cast a new steam chest. They told Colville, 'Your steam chest is in hand & you may be assured there shall not be a moments time lost in finishing & forwarding it to you.'[99] It was not until March 1771 that Colville wrote to Watt asking him to send John Gardner because John Parker was afraid to install the steam chest by himself.[100] The first Boulton and Watt pumping engine in Scotland was erected for Peter Colville at Torryburn in 1776, which shows how important a customer he was.

We do not know the terms of employment of John Gardner, who became Watt's senior journeyman, but we do know that John Gardner and Murray Osborn helped Watt with his own engine at Kinneil. It is probable that Watt retained his more skilled workmen for a time and then employed them on a contractual basis for specific jobs. This was the way he employed some of his first engine erectors after he had moved to Birmingham, and once again we see Watt following a practice in his later steam engine business which he had already established in Scotland.

During the autumn of 1770, Watt installed an engine for Robert Montgomerie of Ayr. While a new engine house was being built, some parts of an old engine were reused. Erection must have commenced that

October, for Montgomerie wrote on Christmas Day,

> I have the pleasure to inform you that on Friday 21s currt, our Engine was sett agoing to the utmost extent of our hopes, she goes admirably well & we are proceeding with our sinking ... Our man John Scot I believe will go mad with Joy at his success, our Engine is going so far superior to that of our Neighbours, I mean Mr Beaumonts, I must own he [the engine] hath surpast my expectations both as to time & workmanship.[101]

The success of this engine came after Watt had had to abandon his own at Kinneil. Presumably knowledge of his reputation must have spread to Black as well as Roebuck in Scotland, but did it spread further afield into England and help to convince Small and Boulton that he was worth supporting?

The last Newcomen engine erected by Watt in Scotland was one for Lord Cochrane at Culross. In February 1772, his lordship had consulted Watt about building a tide mill for pumping water from his mines, but Watt thought the scheme impracticable.[102] The request made in January 1773 that Watt should install a steam engine was countermanded very quickly because Lord Cochrane thought that he had found a second-hand engine at Dysart. However, this one proved to be unsuitable, so that, at the beginning of February, Lord Cochrane ordered Watt to build an engine with a cylinder 40 inches in diameter. The parts were supplied by the Carron Company to Watt's drawings. While the Carron Company queried some of the details on the drawings, they said, 'The plans are very distinct'.[103] No doubt hastened by a letter from Lord Cochrane himself, the parts were ready for Watt's inspection at Carron on 12 June 1773.[104]

What is evident from Lord Cochrane's letters is that Watt was acting in the capacity of a modern consulting engineer. He was to prepare the drawings and pass them on to the respective firms to make the parts. Watt had other suppliers in addition to Carron for copper pipe work etc. Watt was also to prepare instructions for other workmen to carry out their respective functions. While Lord Cochrane was going to be away,

> The charge of carrying on the works in my Absence will fall to Smeaton [presumably not John Smeaton]. Some Instructions on your Part may help to make him both more expert in Carrying on the intended works and more trustworthy in my Absence. Please to send by bearer Directions for my Blacksmith as to the Boiler.[105]

Do we find a parallel here with the 'Directions for Erecting and Working the Engines' which Boulton and Watt drew up for their engine erectors in 1779? We know that Watt did not handle any of the financial transactions necessary to purchase the parts and erect this engine. Lord Cochrane expected that his engine would be running during the early part of August but the Carron Company had to threaten him with court action to obtain payment for what it had supplied.[106]

By the time he moved to Birmingham, Watt had gained considerable experience in designing, erecting and maintaining the Newcomen type of engine. We have seen how Watt, while he retained a small private workshop, moved away from having his own factory and employing his own workmen. His experience in both his civil and mechanical engineering practice (to use modern terminology where Watt would have seen no separation) showed that he could act as a consulting engineer. He had no capital tied up either in productive capacity or in stock. The purchase of the parts and materials was paid for by the customer, which entailed Watt himself in no risk. While he may have employed a clerk, it seems that latterly even the erection of engines was done on a contractual basis so that Watt could avoid supervising workmen.

What we can see here is the beginning of the way Watt would work in Birmingham. Boulton and Watt provided the drawings for their engines. Most of the parts, particularly the cylinders, were ordered from outside suppliers like Wilkinson. Only the regulator valves were produced at Soho. When the parts had been brought together on site, an engine erector was sent, together with printed instructions about how to erect the engine. In this way, the early practice of Boulton and Watt was based mostly on Watt's previous experience in Scotland. Why tie up capital in plant, stock and workmen, especially when equipment like cylinder boring machines might have stood idle for long periods, when you could build engines successfully on a consultancy basis? Does this explain why Boulton's vision of a steam engine manufactory on the banks of a Birmingham canal building engines for the whole world was delayed for 20 years?

Notes and References

1. Birmingham Central Library, James Watt Papers (hereafter JWP), 6/19.23, 7 February 1769, from M. Boulton.
2. JWP 4/18.1, 6 April 1801, from R. Barclay.
3. G. Williamson, *Memorials of the Lineage, Early Life, Education, and Development of the Genius of James Watt* (London, 1856), 152–3; S. Smiles, *Lives of the Engineers, The Steam Engine, Boulton & Watt* (London 1878), 15.
4. Smiles, *op. cit.* (3), 13.
5. Williamson, *op. cit.* (3), 136.
6. JWP C4/B27.
7. JWP C4/B33.
8. Smiles, *op. cit.* (3), 23.
9. D.J. Bryden, 'James Watt, Merchant: The Glasgow Years, 1754–1774', in D. Smith (ed.), *Perceptions of Great Engineers, Fact and Fantasy*, (London, 1994), 9–21, p. 10.
10. JWP 4/11.4, 29 July 1757, to J. Watt senior.
11. JWP 4/11.160, 11 June 1754, to J. Watt senior.
12. JWP 4/11.161, 'A List of James Watt's Clothes Taken to Glasgow', no date.
13. JWP 6/46.34, 19 June 1755, to J. Watt senior.
14. JWP 6/46.30, 21 July 1755, to J. Watt senior.
15. JWP 6/46.6, 19 June 1756, to J. Watt senior.
16. *Ibid.*
17. Birmingham Central Library, Boulton & Watt Collection, MIII, 3/1, Waste Book, 3 January 1757 to May 1763.
18. G. L'E. Turner, *Antique Scientific Instruments* (Poole, 1980), 34.
19. JWP 6/46.1, 19 August 1756, to J. Watt senior.
20. JWP 4/11.1, 27 September and 2, 25 October 1756, to J. Watt senior.
21. B. & W. Coll., MIII, 3/1, 3 January 1757.
22. *Ibid.*, 28 May 1757 etc.

23. *Ibid.*, 2 August 1757.
24. JWP 4/11.9, 10 October 1757, to J. Watt senior.
25. JWP 4/11.21, February 1757 (actually 1758).
26. JWP 4/11.41, 15 September 1758.
27. B. & W. Coll., MIII, 3/1, 18 May 1759.
28. H.W. Dickinson, *James Watt, Craftsman & Engineer* (Cambridge, 1935), 27.
29. B. & W. Coll., MIII, 3/1, August 1759.
30. B. & W. Coll., MIII, 3/4, 7 October 1759.
31. *Ibid.*, 4 October 1760.
32. *Ibid.*, 13 December 1760.
33. *Ibid.*, 10 June 1763.
34. *Ibid.*, 7 March 1764.
35. E. Robinson and D. McKie, *Partners in Science, Letters of James Watt & Joseph Black* (London, 1970), 8, 10 January 1768.
36. Dickinson, *op. cit.* (28), 28.
37. *Ibid.*, 29.
38. JWP C1/1, 69.
39. D.J. Bryden, *Scottish Scientific Instrument-Makers, 1600–1900* (Edinburgh, 1972), 164–71.
40. I am indebted to Michael Wright at the Science Museum, London, for his help with looking at Watt's Garret Workshop.
41. JWP C12/1.33, June 1763.
42. B. & W. Coll., MIII, 3/4, August 1763.
43. H.W. Dickinson and R. Jenkins, *James Watt and the Steam Engine* (1927, reprint Ashbourne, 1981), 19.
44. JWP 4/32.31, 5 December 1765, to J. Roebuck; earlier dates for Craig's death have been suggested but are incorrect.
45. JWP 4/11.134, 23 January 1766, to J. Watt senior.
46. Dickinson, *op. cit.* (28), 29.
47. Robinson and McKie, *op. cit.* (35), 8, 15 February 1768.
48. Dickinson and Jenkins, *op. cit.* (43), 26.
49. Dickinson, *op. cit.* (28), 44.
50. JWP 4/4.18, 4 May 1767, from Mrs M. Watt.
51. JWP 4/7.28, 9 February 1767, from J. Watt senior.
52. Birmingham Central Library, Boulton 348.20, 9 September 1770, to W. Small.
53. *Ibid.*
54. Boulton 348.24, 7 November 1772, to W. Small.
55. Boulton 348.25, 24 November 1772, to W. Small.
56. JWP 3/43, 7 and 21 July 1770, from J. Naismith.
57. JWP 3/43, 'Regulations for the Undertaker of Hamilton Bridge', no date.
58. JWP 4/19.15, 7 July 1771 and 16, 9 August 1771, from W. Mather.
59. JWP 4/50.7, 17 August 1771, to Lord Provost of Glasgow.
60. R.H. Campbell, *Carron Company, Founded 1759* (Edinburgh, 1961), 7.
61. Scottish Record Office (hereafter SRO), GD58/2/1/1, 162, 31 January 1765.
62. SRO, GD58/6/1/6, 12, 1 September 1766, to F. Garbett (?).
63. SRO, GD58/6/1/11, 69, 23 August 1770, to C. MacDowall.
64. SRO, GD58/6/1/12, 298, 30 December 1772, to P. Buchanan.
65. Smiles, *op. cit.* (3), 98.
66. B. & W. Coll., MI, 1/15, 12 June 1770.
67. JWP 4/32.36; see also J.P. Muirhead, *The Origins and Progress of the Mechanical Inventions of James Watt, Vol. I* (London, 1854), 7.
68. SRO, GD58/2/3/1, 89, June 1770.
69. JWP 4/32.106, 4 November 1769, from J. Roebuck junior.
70. JWP 4/32.14, 15 April 1770, from Dr J. Roebuck.
71. B. & W. Col., MI, 1/15, 4 June 1770 and JWP 4/32.102, from Dr J. Roebuck.
72. JWP 4/32.101, 5 June 1770, from J. Roebuck junior and B. & W. Coll., MI, 1/15, 14 and 18 September 1770.
73. JWP 4/19.18, 21 April 1772, from Orrs & Wilson.

74. B. & W. Coll., MI, 2/3. Cash Memorandum Book, April 1765 to January 1767.
75. J. Lindsay, *The Canals of Scotland* (Newton Abbot, 1968), 15–16.
76. Dickinson, *op. cit.* (28), 44.
77. Lindsay, *op. cit.* (75), 20.
78. SRO, GD58/6/1/7, 495, 17 October 1768, to R. Mackell.
79. SRO, GD58/6/1/5, 75, 24 August 1765, to R. Mackell.
80. SRO, GD58/6/1/5, 187, 2 December 1765, to J. Monro.
81. JWP 4/52.1, 5 December 1765.
82. *Ibid.*
83. JWP 4/52.3, no date.
84. Muirhead, *op. cit.* (67), *Vol. III*, frontispiece.
85. *A Catalogue of the Civil and Mechanical Engineering Designs, 1741–1792, of John Smeaton, FRS* (London, 1950), 58, Newcomen Society Extra Publication No. 5.
86. J. Farey, *A Treatise on the Steam Engine, Historical, Practical and Descriptive* (London, 1827), 262.
87. Patent 895, 30 June 1768.
88. B. & W. Coll., MIII, 3/6, 59, 11 April 1766.
89. JWP 4/52.4–6.
90. JWP 4/52.6.
91. B. & W. Coll., MI, 2/3, 3 June 1766 etc.
92. JWP 4/52.7.
93. JWP 4/52.8, 15 May 1769, from Lord Kennet.
94. JWP 4/52.9, 21 December 1766, to Mr Dick.
95. B. & W. Coll., MIII, 3/2, September 1768 and JWP 6/28.82, 22 October 1769.
96. SRO, GE58/6/1/6, 173, 23 January 1767 and 198, 16 February 1767, to P. Colville.
97. SRO, GD58/6/1/10, 315, 9 May 1770, to P. Colville.
98. B. & W. Coll., MIII, 3/2, August 1768 and MI, 1/14, October and November 1768.
99. SRO, GD58/6/1/11, 301, November 1770, to P. Colville. See also GD58/6/1/11, 240, 22 October and 241, 23 October 1770 to J. Watt.
100. JWP 4/19.17, 2 March 1771, from P. Colville.
101. JWP 4/19.8, 25 December 1770, from J. Montgomerie.
102. JWP 4/19.24, 23 February 1772, from Lord Cochrane.
103. SRO, GD58/6/1/12, 389, 2 March and 429, 23 March 1773, to J. Watt.
104. SRO, GD58/6/1/12, 380, 23 February and 584, 12 June 1773, to J. Watt.
105. JWP 4/19.32, 14 February 1773, from Lord Cochrane.
106. SRO, GD58/6/1/13, 394, 18 February 1774, to Lord Cochrane.

Technological Advance in the Manufacture of Chemicals: The Case of Cyanide, 1888–1930

ALAN L. LOUGHEED

INTRODUCTION

When there is little demand for a commodity, its production may be rudimentary and relatively costly. If, however, for some reason such as the discovery of an important new use for the product, there is a sharp and expanding increase in its demand, then research may be stimulated to discover new production methods which would reduce the price of the product and widen its usage. Competition may also provide an incentive for greater research effort. The manufacture of potassium (and sodium) cyanide appears to fit such predictions.

Until 1887, potassium cyanide was used only as a raw material and only by electroplaters, photographers and pharmaceutical producers. Germany made most of it, production being under the control of the *Deutsche Gold und Silber Scheideanstalt (DEGUSSA)*, the selling agent for all producers. Total output in 1886 was around 200 tonnes.[1]

At that time, potassium cyanide was made from potassium ferrocyanide (also called yellow prussiate of potash), $(NC)_6FeK_4 . 3H_2O$. The latter was produced industrially from animal refuse (hides, horn clippings, old shoes and blood solids), potassium carbonate and iron filings or borings. By the fusing of these materials at a red heat, the prussiate passed into solution, leaving a black residue. The clarified solution was then heated for a time and, when it was cooled, deposits of lemon-yellow quadratic crystals of ferrocyanide appeared. These were purified by recrystallation. Potassium cyanide was then produced commercially by fusing a mixture of eight parts anhydrous prussiate and three parts anhydrous potassium carbonate, allowing the reaction

$$(NC)_6FeK_4 + K_2CO_3 = CO_2 + Fe + 5KCN + KNCO$$

to complete itself. The iron (Fe) was settled and the clear fuse was decanted off. Fused potassium cyanide (KCN), when cooled, was a milky-white stone-like solid.[2] This crude process of making prussiate was sufficient to satisfy the low demand for it. Any large increase in demand, however, could cause the prices of the ingredients to rise because of their rapidly growing scarcity, and thus lead to an increase in the price of the product. Clearly, such a price rise would stimulate research for a new method of producing prussiate and, eventually, a prussiate-free method of making cyanide.

In 1887, John Stewart MacArthur, the technical manager of the Cassel Gold Extraction Company (henceforth Cassel) in Glasgow, developed the cyanide process for extracting gold from its ores and tailings. Of all the chemical substances tested by MacArthur for extracting gold without affecting the base metals with which it may have been combined, potassium cyanide was the only useful one and it acted most efficiently in a very dilute form. This discovery was to transform the gold extracting industry in many parts of the world, for its cheapness and efficiency meant not only that a very high rate of gold extraction was now possible but also that very low grade ores could be treated profitably for the first time. The advances in gold production after 1890 were to a large extent owing to this discovery and its use on the world's goldfields.[3]

Naturally, such a technological advance affected the production of potassium cyanide. At first, Cassel purchased its requirements through the trade, but when the amounts required increased, the company's attention turned towards its manufacture.

THE EARLY YEARS

It is useful to follow the development of cyanide usage by tracing the early activities of the prime mover at the time, the Cassel Company. Of necessity the cyanide process had to be taken to the world's goldfields overseas and put into use as soon as possible. After the Cassel board sent teams of cyaniders to Queensland and New Zealand in 1888, the need for more potassium cyanide became urgent. In September, 20 tons of prussiate[4] were purchased and premises were found in Glenpark Street, Dennistoun, Glasgow, to convert the prussiate into cyanide. Within 2 months production had begun. On 6 December, the first order was received from New Zealand. The price was set at 1*s* 9*d* a pound for 100 per cent purity.[5] Altogether, around 30 tons were produced at the Cassel works in 1888 and 50 tons in 1889. Most of this cyanide was used in New Zealand and at the Cassel site near Ravenswood in Queensland.

By mid-1890, cyanide was being used for gold extraction in the Transvaal and an order for 50 tons of cyanide was received for delivery over 3 months from October 1890. This large order was to be followed by many more. Cassel's output for 1890 was 200 tons.

By the end of 1891, production at Glenpark Street expanded rapidly to cope with the rising South African demand. Supplies of prussiate were obtained from local and German producers. At the same time, the rapidly rising demand for prussiate led to a price increase of 50 per cent between 1888 and the end of 1891. Although economies of scale undoubtedly occurred as cyanide production rose, the increased price of prussiate resulting from the increased demand for it and the growing scarcity of its ingredients prevented the price of potassium cyanide from falling at this time. Clearly, other methods of production had to be found.

THE EARLY BRITISH–GERMAN CONNECTION

The development of the MacArthur-Forrest process was a boon to the German cyanide industry. By 1886, an international ferrocyanide cartel had been formed by the German manufacturers and at least two British firms to control sales, production and the price of prussiate.[6] DEGUSSA acted as the selling agent for both the German prussiate and cyanide producers. When cyanide was introduced into gold extraction, the German producers adopted a wait-and-see attitude before interfering with Cassel's position, but not for long. By the end of 1891, the rapidly rising demand for the product in the Transvaal could not be ignored by the German firms. The cyanide process had proved to be successful.

Cassel's attempts to force the mining companies using the new process to buy its cyanide were unacceptable to the German producers. Such attempts could not be sustained in view of the price cutting and discounting available to Cassel's competitors. As a result, some of its market was captured, not only by the Germans but also by new British producers. Early approaches were made to Cassel by DEGUSSA for the formation of an international cyanide makers' cartel to regulate sales and prices, but the move was somewhat premature.[7] On the contrary, by 1894, the stage was being set for a major cyanide price war. However, the Cassel board was well aware of the need for production methods to change in order to reduce costs, so that the company could be placed on a sounder competitive footing *vis-à-vis* its rivals.

THE FIRST EXTENSION OF CYANIDE TECHNOLOGY

There were therefore very good reasons for an entirely new approach to cyanide manufacture as the shortages of raw materials for the manufacture of prussiate grew, forcing up the costs of the cyanide makers. Early in January 1892, George Thomas Beilby approached Cassel with an entirely new method of producing cyanide cheaply. The attractiveness of the method ensured that every encouragement was given to Beilby, who was requested to proceed with further experiments at the company's expense.

This process involved 'the passing of ammonia over or through a mixture of anhydrous alkali with alkaline cyanide and finely-divided carbon at a temperature sufficiently high enough to keep the mixture fluid and to decompose the ammonia'.[8] Plans were supplied for an appropriate plant. In practice, Beilby mixed about 20 per cent potassium cyanide, 20–25 per cent carbon and 55–60 per cent carbonate of potassium and passed ammonia through the mixture. But it was to take another fourteen months before this new cyanide could be fed into the market, and several more before the company could dispense altogether with its 'prussiate' method.

A Cassel rejection of German prussiate in October 1892 was not well received by the German parties, but further discussions with DEGUSSA's London agents produced a new agreement committing Cassel to the purchase of 10 tons of prussiate a week from German sources throughout 1893. This stop-gap measure, essential to the company's operations until Beilby's cyanide came on stream and to the maintenance of as much secrecy as possible regarding Beilby's progress towards that end, was highly unsatisfactory for the board. In May 1893, it received information that DEGUSSA was determined to increase German competition in South Africa, and that the company had begun experiments on another ammonia-based production method.[9] The Germans were emulating Cassel's approach and the Cassel board recognized the urgency of bringing Beilby's method into production, as it could give the company the competitive edge essential in a price war.

Cassel obtained new premises in Shuna Street, Maryhill (in north Glasgow), and a supply of ammonia was purchased. Even so, larger orders for cyanide from Africa led to an increase in the number of furnaces installed at Cathcart from 12 to 18, and in September 1893 more prussiate was purchased from local producers, while supplies were still being received from Germany under the 1892 agreement. In December 1893 the first 10 tons of Beilby's cyanide were sent to Australia. The next stage of the cyanide saga had begun. It does appear that the uncertainty in Cassel regarding the stage reached by German manufacturers in the development of their synthetic methods played an important role in speeding up the introduction of the company's new, revolutionary technology. Yet it was undoubtedly clear that the German producers knew what was happening at Cassel. As noted above, Beilby's patent had been published, and a little industrial spying would have yielded all the information the German rivals required concerning Beilby's process. Nevertheless, the first use of the new technology was made by the innovating firm, which hoped to be able to recover most of its developmental costs before its competitors caught up with it. They were not far behind.

Cassel's commitment to the old technique persisted until the Beilby cyanide production could be expanded and until costs could be reduced. Early in 1894, the increasing price of prussiate forced a rise in the price of cyanide to 1*s* 2*d* for 72 per cent quality. In the last three months of 1893, Beilby's costs had been reduced to average 8.44 pence a pound for 72 per cent purity. Considering the asking price for 'prussiate' cyanide, his process allowed Cassel a temporary but sizeable competitive advantage. Gradually throughout 1894 the new brand was fed into the market. By May of that year, the Maryhill plant was producing around 8 tons a week and it was soon decided that the works should be extended. It was not long before Beilby's process was the only one used by the firm.

Meanwhile, in the second half of 1894, a new attempt by DEGUSSA to form an international cartel met with a 'friendly' reply of refusal from Cassel and, despite all German attempts to influence the Scottish company, such approaches at that time were resisted.

In the closing months of 1894, the Cassel board prepared the company for a period of aggressive competition in South Africa. Attempts were made at Cathcart to reduce the costs of making 'prussiate' cyanide and in November it was reported that a considerable reduction to around one shilling a pound for 100 per cent had been achieved. By then, Beilby had increased production at Maryhill to about 15 tons a week, with an appreciable fall in his production costs.

INCREASING COMPETITION AND THE PRICE WAR OF 1896–7

During 1895, several British firms entered the cyanide industry. Entry was relatively easy, partly because of the large increase in the number of patents registered for producing potassium cyanide,[10] and partly because of the fall in the price of prussiate when the large cyanide producers moved into ammonia-based production. The new British firms included British Cyanides, Johnson Brothers, May and Baker, the London Gas, Light, and Coke Company and the North British Chemicals Syndicate. These firms concentrated largely on prussiate technology using the wastes of gasworks for producing the prussiate.[11] Thus, another path to the production of cyanide had been found, with more to follow. The prussiate approach was also enhanced by the rapid fall in the price of the chemical from 10.5 pence a pound at the beginning of 1895 eventually to 5.5 pence in June 1897, a result of the decline in demand from the larger firms and the increasing supply from the gasworks. Finally, in 1895, the new producers shared in the profits of cyanide production by undercutting the prices of their larger rivals, which was possible because of the high profit margins of the latter.

By April 1896, however, the large discounts being offered by Cassel's competitors to its South African clients led the firm to lower its price from around 16.5 pence to around 11–12 pence a pound for 98 per cent quality during the remainder of 1896. In the first half of 1897, further reductions occurred until 9 pence a pound was recorded. Even so, price discrimination was common. The main battlefield for the cyanide producers was in South Africa.

To a large extent, the price war was a necessary phase in the development of an industry easy to enter and comprising both large and small producers, using different productive processes. In addition, the stimulation of research into new production processes inevitably led to lower prices, the benefits of which were to be shared by many.

The price war carried into 1897. Most small producers weathered it even if their profits were greatly squeezed. In the prevailing industrial climate, cartel arrangements seemed inevitable and such a combination was created towards the end of June 1897, covering all British and German producers. By this time, the British producers included Cassel,

May and Baker (under the control of J. A. Barnicott[12] and receiving supplies of prussiate from the London Gas, Light, and Coke Company, supervised by W. G. Blagden), British Cyanides and the North British Chemical Syndicate (the last-mentioned going into liquidation in April 1900, and its position being taken by Blagden and Waugh, formed early in 1899). In Germany, in addition to DEGUSSA, the major producer was Vorster and Grüneberg at Stassfurt,[13] but other firms were to join these.[14] In the United States, an importer of German cyanide, the firm of Roessler and Hasslacher[15] began to produce 98 per cent pure potassium cyanide at a factory at Perth Amboy in New Jersey in 1894, obtaining its prussiate from local producers and from the German ring. Because of the high tariff on imported cyanide, this relatively high cost producer obtained a virtual monopoly in the United States market until lower tariff rates in 1897 made its production uncompetitive for a time.

FURTHER DEVELOPMENT IN CYANIDE MANUFACTURING TECHNOLOGY

Several other developments in the 1890s affected the attitudes of the major producers towards their manufacturing techniques. DEGUSSA controlled the Castner process for making sodium cyanide (NaCN) and was perfecting this process at Niagara in the United States.[16] It was also considering the possibility of working the process at Oldbury in England.[17]

Previously, in 1891, DEGUSSA had entered into an agreement with the Aluminium Company to obtain its future requirements of sodium from that source. The Aluminium Company had purchased the rights to the Castner process of manufacturing sodium, patented in 1886, while DEGUSSA later also obtained the right to use the Castner process for producing sodium cyanide, patented in 1894. This method involved 'the passing of ammonia over sodium heated in an iron retort yielding sodamide, which is brought into contact with red-hot charcoal so as to be converted into sodium cyanide'.[18] Towards the end of the 1890s, DEGUSSA was attempting to develop Castner's process commercially but initially the end product contained both cyanide salts.[19]

In the United States, DEGUSSA, the Aluminium Company and Roessler and Hasslacher had pooled resources in 1895 to set up the Niagara Electrochemical Company to manufacture metallic sodium using the hydroelectric power provided by the Niagara Falls.[20] DEGUSSA and the Aluminium Company established the Electro-Chemische Fabrik Natrium in 1897 to manufacture sodium using cheap hydro-electricity at Rheinfelden in southern Germany.[21] Nevertheless, despite the intentions of DEGUSSA to move into the production of sodium cyanide, no pure sodium cyanide appeared in the market before 1902.

In 1895, the Castner-Kellner Alkali Company was formed in London to use the combined patents of Hamilton Castner and Karl Kellner for the manufacture of caustic soda and bleaching powder by means of the electrolytic decomposition of salt or alkaline salts, under arrangements with the Aluminium Company, the holder of the Castner patents, and

Solvay et Cie of Brussels, holders of the Kellner patents. Works were established at Weston Point near Liverpool. In 1900, this company absorbed the Aluminium Company and transferred its works from Oldbury to its Weston Point property, where electricity was comparatively cheap and where raw materials for sodium manufacture were readily available.[22]

In Germany, two other innovations occurred in the cyanide industry in the 1890s. First, in addition to the low-cost production of prussiate from the spent oxides of gasworks, used by a number of firms to produce cyanide, another process involving the direct use of gas was developed by Dr Julius Bueb of the German Continental Gas Company at its Dessau gasworks.[23] This method proved so successful that a company, the Chemische Fabrik Residua, was formed in 1898 to exploit Bueb's innovation. It maintained production of cyanide until 1913, when it was liquidated because the method of producing gas had changed.

The other German innovation of the 1890s was the use of beet sugar residues from the Dessau molasses refinery. Such residues were called *Schlempe*, a name given to the type of cyanide produced.[24] This method was devised by Bueb and Reichardt in Dessau. In March 1897, DEGUSSA influenced the Dessau company to restrict its output temporarily to 300 tons a year. In the following year, a new company, the Chemische Fabrik Schlempe, was set up to take over the manufacture of cyanide by this process. A small factory was established at Taucha near Leipzig, where NaCN was produced from the *Schlempe* obtained from the sugar refineries at Dessau, Hildesheim, Oschersleben, Rositz and Gross-Mochbern. Although the Taucha works were extended in 1908, the amount of cyanide produced there was always controlled by DEGUSSA.[25]

The first pure sodium cyanide (130 per cent quality of KCN)[26] was produced commercially by a process developed in Scotland by Dr William Readman in 1894. It involved the electrical production of barium cyanide, which was converted by a wet chemical process into sodium cyanide.[27] Readman set up the Scottish Cyanide Company in 1898 and erected a plant at Leven on the Fife coalfield to produce NaCN at a reduced commercial cost.[28] But Readman was very unlucky. Just as he began production, cyanide sales became depressed by the waging of the Boer War. In addition, sodium cyanide in a pure form had still to be accepted by the mining industry as a whole. The company's stocks could not be sold and it was liquidated in 1907.[29]

It was not until March 1899 that the Cassel Company decided to change its product from potassium cyanide to sodium cyanide, using a process similar to Beilby's. To this end, the Maryhill plant was enlarged and an agreement for a continuous supply of sodium was negotiated with the Aluminium Company, an agreement which was continued by the Castner-Kellner Company in 1900. The two companies also entered into a profit-sharing arrangement.

Given all these changes, it is obvious that cyanide could be produced by several very different methods. Moreover, the industry was changing despite the restrictive aspects of the cartel arrangements. Considerable sums of money were being used on research and development and on the construction of buildings and plant to adopt the new methods of production as viable commercial propositions, all undertaken at a time when total cyanide output had to be limited. Of the various methods used to make cyanide, it does appear that the *Schlempe* method was the cheapest, as sodium cyanide emerged as a joint product with sulphate of ammonia, such that costs of production could be allocated between two products.

The turn of the century was a time in which several innovations were occurring and the major producers were compelled to go with the tide. Offsetting the declining demand emanating from South Africa, the increased gold production in other countries portended a steeply rising trend in cyanide demand in the 1900s. According to Beilby, the consumption of cyanide in gold extraction throughout the world was 2600 tons in 1897 and 5500 tons in 1902.[30]

By 1902, the changeover to sodium cyanide had begun. The Castner process was adopted in the major cyanide-producing countries and, as noted above, the leading companies acquired interests in the sodium factories that were established. In addition to the new sodium works at Rheinfelden and Niagara Falls, another joint venture was under way by 1901. In France the Société d'Electrochimie de Paris had already begun to make potassium cyanide in a small way and in 1900 it had obtained a licence from DEGUSSA to manufacture sodium cyanide at its works at Les Clavaux. It also planned to set up a factory to produce sodium at Martigny in Switzerland. DEGUSSA and Cassel assisted the company and set its quota under the cartel at 400 tons a year.[31]

STATE OF THE CYANIDE INDUSTRY, 1904–14

By 1904, sodium cyanide was becoming the major form in use, even though for some time there was some resistance to the change. Moreover, because the United States tariff on sodium cyanide was twice that on potassium cyanide (and the mixed variety), a mixed cyanide was imported mainly from Germany for much of the period. The United States Board of General Appraisers investigated 80 samples of the imported double cyanide and found the sodium content to be as high as 22 per cent. Nevertheless, it was decided to levy the lower rate on the mixed salts 'on the general ground that the ordinary commercial potassium cyanide often contained impurities, especially sodium compounds'.[32] Such was the situation until 1912, when all cyanide salts were classified under the higher rate of duty. Although imports of the mixed salt guaranteed to be equivalent to 100 per cent potassium cyanide continued to be imported into the United States for many years, the swing towards pure sodium cyanide gathered pace.[33]

Cyanide users in most countries found that they were favoured not only by the fall in prices in the early years of the decade but also after 1904 because substantial economies were made in the use of more dilute cyanide solutions to obtain the same output. As a result, although cyanide usage expanded rapidly during these years, the increasing sales of cyanide did not match the growth of output of gold and silver obtained from its use.

The major forms in which cyanide was produced included the fused Castner method involving sodium and ammonia, *Schlempe* using beet sugar residues and the Bueb method using gas. By 1913, Germany produced 5000 tons of sodium cyanide (valued at US$1,750,000) and 5000 tons of ammonium sulphate from *Schlempe*. But *Schlempe* production varied from year to year, depending on the size of the beet sugar crop and on the demand for cattle cake. If the latter was high, the beet sugar residues tended to be used for that purpose.[34]

In the United States, Roessler and Hasslacher began its manufacture of NaCN in 1902, obtaining its sodium from the Niagara Electro-Chemical Company. There were other producers; for example, the Davis Chemical Company in Massachusetts was producing small amounts.[35] In the UK cyanide was being used in increasing quantities for other purposes. In 1910, 2000 tons of sodium cyanide went into ore treatment and 2500 tons were used for fumigation, steel hardening and electro-plating. The eradication of pests in the Californian citrus orchards involved the use of hydrocyanic acid.[36]

The Cassel Company was by far the largest cyanide producer in Britain (and the world), and it was one of the only producers without a diversified product mix. The United Alkali Company became a producer at the turn of the century, first experimenting at its Runcorn plant in making potassium cyanide from sulpho-cyanide using the wastes of gas works.[37] It also began producing sodium cyanide using the Raschen process involving carbon disulphide and ammonia.[38] It continued to make and sell small quantities until 1908, when it agreed to cease cyanide manufacture and purchase its requirements from British Cyanides.

During this period, Cassel expanded its output (often selling some to other British producers to fill their quotas) from 1732 tons in 1900 to 6622 tons in 1909. In the latter year, total British exports were 7286 tons. With expansion, certain arrangements were required. First, it was necessary to obtain increasing amounts of each raw material from year to year. Brotherton's of Leeds, the company's supplier of ammonia, was willing to increase its supply (there were close financial connections between the two companies). Second, an increasing supply of sodium was essential at the lowest cost possible. Clearly, electricity was a major component in the production of sodium, as the text above indicates. In 1905, Cassel and Castner-Kellner came to a joint agreement to erect a sodium factory at Wallsend-on-Tyne, where electricity was cheaper than at the Weston Point Castner-Kellner plant. Construction of the plant commenced in 1906 and production was well under way by 1907. Many extensions of the plant were required over the years to 1914.

Despite these arrangements, Cassel strengthened its sources of supply of sodium by contracting to take up to 200 tons a year from 1908 to 1916 from the Vadheim Elektrochemiske Fabriker company in Norway.[39]

WORLD WAR I AND ITS AFTERMATH

The outbreak of World War I in 1914 placed the Cassel Company in a dominant position in the cyanide industry, the German supplies to the goldfields evaporating overnight. This position was held until the end of hostilities and the company's output expanded rapidly. Meanwhile, the uses of cyanide were also changing, especially in the United States, where, by 1920, total consumption had reached some 8000 tons a year, 2000 for gold and silver extraction, 2400 for electroplating, 1600 for the hardening of steel, 1200 for fumigation of orchards, ships etc. and 800 for dye, colour and other industries.[40] Also by then the established cyanide producers were challenged by a new formidable competitor, the American Cyanamid Company.

In 1910, this company had set up a plant on the Canadian side of Niagara Falls to use the cheap Canadian electricity, and began to manufacture the fertilizer calcium cyanamide. For several years, it had been known that calcium cyanamide could be transformed into cyanide and, in 1917, the company began to move in this direction to take advantage of the expanding cyanide market in the United States. Its process for manufacturing sodium cyanide consisted of mixing calcium cyanamide and common salt, and melting the mixture in a single-phase electric furnace. The operation was continuous. The final product, the Aero brand, was low-grade sodium cyanide of 36 to 46 per cent potassium cyanide equivalence. It dissolved easily but had to be agitated.[41] The company also produced cheap low-grade potassium cyanide.

Comparisons of the new brand and high-quality cyanide show mixed results. Despite the impurities in the Aero brand, tests by John G. Kirchen at a mill in Nevada, using the 36 per cent Aero and a 97 per cent other brand cyanide, showed approximately the same extraction rate, between 92 and 93 per cent, but that the consumption of cyanide per ton and per ounce of gold extracted slightly favoured the high-grade variety. A more extensive test slightly favoured the Aero brand.[42] In Canada in 1922, on the other hand, the metallurgists of the rich Porcupine and Kirkland Lakes districts of Ontario still preferred high-grade cyanide,[43] while, in the Cobalt district, some use was made of the Aero brand, although it was mainly discarded because it was rich in lime, and thus not favoured there because soda ash had to be obtained to neutralize it.[44] It also proved unsuitable for use in electroplating, steel hardening and rust proofing, where virtually impurity-free cyanide was still required.

The entry of this cheap cyanide into the market led temporarily to reduced sales by the other producers, all of which attempted to find new production methods but in the meantime let their profit margins fall so that they could still record some increases in sales over time. The Cassel Company, for example, began to produce a low-quality (72 per cent) variety and attempted to compete with it. Nevertheless, throughout the 1920s, the international cyanide convention was maintained and all cyanide producers were able to sell their quotas, which were altered occasionally to meet the demands of American Cyanamid. After 1925, surplus stocks of *Schlempe* cyanide had to be run down over a 3-year period, each cartel member accepting lower quotas in the years involved.

The 1920s also became a decade of technological experiment with the goal of increasing competitiveness *vis-à-vis* American Cyanamid. At Cassel, the Ewan method proved to be an advance on Beilby's, but Ewan's main problem was how to maintain a uniform quality (a problem ignored by American Cyanamid). In Germany there were three main areas of research. DEGUSSA registered two new patents. The first was for a method of obtaining hydrocyanic acid by the reaction of ammonia and carbon monoxide (patented in Britain in 1923), while the second envisaged the manufacture of hydrogen cyanide by the interaction of ammonia and carbon monoxide with various metal oxides as catalysts (British Patent No. 25147 of 1924).[45] The Badische Anilin und Soda Fabrik (BASF) was contemplating entry into cyanide production. In 1923 it patented a process of producing hydrocyanic acid from carbon monoxide and ammonia using alumina as a catalyst (BP No. 17467 of 1923) while, in 1924, it demonstrated an interest in a process of manufacturing NaCN from gases containing hydrogen cyanide and carbon dioxide with or without ammonia (BP No. 5345 of 1924). In addition, I. G. Farbenindustrie of Ludwigshafen began to make hydrogen cyanide by the catalytic dehydration of formamide, and this was dissolved in a caustic soda solution, producing another solution from which sodium cyanide was recovered. By 1926 it was in a position to produce 3000 tons of 92 per cent a year.[46] In France, the Fabriques de Produits Chimique de Thann et de Mulhouse obtained patent rights to a process of producing cyanide from cyanamide of lime by reaction with calcium carbide and sodium carbonate, thus using the Aero brand as a point of departure.[47] In the United States, the Du Pont de Nemours Explosives Company of Wilmington, Delaware, was attempting to improve its new sodium cyanide process of heating a mixture of soda ash, carbon and catalysers and passing nitrogen through the mixture under pressure.[48]

The publication of all these endeavours shows the extent to which industrial chemists were attempting to widen the product mixes of their respective employers. But many processes proved incapable of competing commercially, at least in the short run.

Cassel was absorbed by ICI in 1927 and production of cyanide was transferred from Maryhill to Billingham South, where both sodium and cyanide were produced thereafter. Roessler and Hasslacher was taken over by Du Pont in 1930. Both the absorbed companies retained their names for trading purposes for several years. By 1930, the world's cyanide production was controlled by four large companies: American Cyanamid, Du Pont, ICI and DEGUSSA. I. G. Farben and CIBA (a new entrant in Germany) were both limited by the cartel arrangements to small quotas which were sold to the big four; *Schlempe* cyanide was still produced at Dessau and Kolin[49] and controlled by DEGUSSA, and the French Electrochimie produced a limited annual amount, its output being used in France and the United States for industrial purposes.[50]

In 1931, Cassel reached agreement with Roessler and Hasslacher, allowing the latter to manufacture cyanide using the Ewan process for the manufacture of sodium cyanide from sodium, charcoal and atmospheric nitrogen,[51] but Cassel chemists were progressing towards an improved process of making sodium cyanide from methane and ammonia, first by producing hydrocyanic acid by high-temperature catalytic reaction of methane with ammonia and air, and second by adding this acid to a caustic soda solution to produce the sodium cyanide.

CONCLUDING REMARKS

Such was the state of play in cyanide production in the 1930s, the market for which was dominated at the time by American Cyanamid and Cassel, which together accounted for around 75 per cent of the world market. Roessler and Hasslacher accounted for around 13 per cent and the German and French producers shared most of the remainder almost equally. The anticipations of some of the new German manufacturing methods did not come to fruition and they could not compete with the new Cassel–ICI processes for high-quality cyanide nor with the Aero brand for the low-quality cyanide.

It is clear, however, that once a sizeable market had developed for the product, research and development became a feature of the growing cyanide industry; so much so that numerous methods were soon developed and used in production. But research could not be slackened and Cassel for one had a research programme in operation continuously from the late 1880s. Indeed, the company was one of the few one-product firms operating in the industry, and its very existence depended upon maintaining a leading position in research and development. Another feature of the chemistry of the commodity was that up to World War I entry into the industry was easy through the adoption of one of a number of readily available processes. Even so, it was recognized that an international cartel was required to protect the research investments of the major participants. Such arrangements survived throughout the 1930s despite the encroachment of the Aero brand.

Finally, there is little evidence in this section of the British chemical industry of the general malaise and the growing technological backwardness that is often referred to in comparisons between British and Continental and American industrial sectors between the 1880s and the 1920s. Some sections of British industry were not only holding their own but acting as technological leaders.[52]

Notes and References

1. See Anon., 'Deutsche Gold und Silber Scheideanstalt, 1873–1923', *Die Chemische Industrie*, 1923, 46: 99–104, p. 101.

2. For this information, see *Encyclopedia Britannica*, 9th edn (London, 1883), 20–2. Note, however, that the low grade KCN (70–75 per cent pure), first to be used on the goldfields, was greyish in colour.

3. See G.T. Beilby, 'Ten Years' Progress of the Cyanide Process for the Extraction of Gold', *Journal of Society of Chemical Industry* (henceforth *J. Soc. Chem. Ind.*), 1898, 17: 130–8. Note also that the chlorination process had already been developed, allowing very high

rates of extraction, but its use was limited on most goldfields because it was not a relatively cheap method of extraction. In addition, the traditional (mercury) amalgamation process was not capable of extracting much gold from refractory ores. See, also, my paper, 'The Cassel Company, Cyanide, and the Gold Mining Industry, 1887–1927', delivered at the International Mining History Conference, Melbourne, Australia, August 1985; and D.I. Harvie, 'John MacArthur: Pioneer Gold and Radium Refiner', *Endeavour* (new series), 1989, 13: 179–84, pp. 179–81.

4. 'Prussiate' will generally be used for 'potassium ferrocyanide'.

5. The price varied according to the degree of purity of the product. In the early days, 100 per cent purity (without foreign matter) was rare and 70–75 per cent was common; 72 per cent purity sold at around 1*s* 3*d* a pound in 1889.

6. See 'DEGUSSA', *op. cit.* (1), 101; H. Mayer-Wegelin, 'Marktregelungen in der Chemischen Industrie', *Die Chemische Industrie* (new series) 1952, 4: 745–8, p. 747.

7. Details of these relations are found in the *Board Minutes* of the Cassel Company (henceforth *CBM*), held in the Cheshire Record Office in Chester.

8. See *J. Soc. Chem. Ind.*, 1892, 11: 747, 1004.

9. It was only a matter of time before DEGUSSA began to follow Cassel in this new direction. Beilby's process was described in some detail in *J. Soc. Chem. Ind.*, 1892, 11: 747, and his connection with Cassel would soon have been discovered. Also, a similar process preceded Beilby's and was outlined in *J. Soc. Chem. Ind.* 1890, 9: 860, under an English patent obtained by H. Grüneberg, H. Fleming and W. Siepermann of Germany, in August 1889. The latter was adopted by Vorster and Grüneberg at Stassfurt in 1892. Beilby's end product was easier to work up into a readily marketable form. See *Thorpe's Dictionary of Applied Chemistry* (London, 1939), *Vol.* III, 809.

10. Beilby noted five patents in 1891, six in 1892, 13 in 1893, 26 in 1894 and 30 in 1895 ('Ten years' progress', *op. cit.* (3), 134).

11. In the early 1890s, several patents were registered for the use of the wastes of gasworks. See, for example, *J. Soc. Chem. Ind.*, 1893, 12: 511, for an outline of that of J. Foulis of Glasgow.

12. May and Baker acquired Johnson Brothers' cyanide business and plant in July 1896. Barnicott came to Battersea with the plant to manage their cyanide production. See J. Slinn, *A History of May and Baker 1834–1984* (Cambridge, 1984), 60–3.

13. Vorster and Grüneberg became the Stassfurter Chemischen Fabrik in 1897 (see Anon., 'DEGUSSA', *op. cit.* (1), 101).

14. See below.

15. A subsidiary of DEGUSSA which held 51 per cent of its shares.

16. See below.

17. At the premises of the Aluminium Company.

18. See Anon., 'Cyanide Supply', *The Mining Magazine*, 1914, xi: 89–90. This resembles Beilby's process if sodium is substituted for potassium in the Beilby specification.

19. See A. Whitby, 'Notes on Commercial Cyanide of Potassium', *Engineering and Mining Journal* (henceforth *EMJ*), 1903, 75: 333. Whitby's experiments showed that the mixed cyanide of 98 per cent potassium cyanide quality allowed extensive adulteration and he found that pure sodium cyanide was more acceptable than the mixed variety.

20. See the report of a paper delivered at the Mining and Metallurgical Society of America in New York, 15 October 1914, by L. Ruhl of Roessler and Hasslacher, *The Mining Magazine*, 1915, xii: 45.

21. See Anon, 'DEGUSSA', *op. cit.* (1), 103.

22. See ICI, *Fifty Years of Progress, the Story of the Castner-Kellner Alkali Company 1895–1945* (Cambridge, 1945), Chapters 3 to 5.

23. See Dr J. Bueb, 'The Production of Cyanides in Gasworks', *EMJ*, 1900, 70: 485–6.

24. C.A. Browne, 'Cyanide from Residue of Sugar Mills', *Mining and Scientific Press*, 1913, 107: 186.

25. See Anon., 'DEGUSSA', *op. cit.* (1), 101–2.

26. Pure sodium cyanide contains about 30 per cent more cyanogen (the active element of the compounds) than pure potassium cyanide: thus the 130 per cent, taking pure potassium cyanide as 100. The other reason for change was the lower cost of producing the sodium salt. See Anon., 'The Two Cyanides', *EMJ*, 1905, 79: 1241.

27. See note in *EMJ*, 1904, 78: 271.
28. See Beilby, 'Ten Years' Progress', *op. cit.* (3), 134.
29. See Readman's obituary, Part 1, *J. Soc. Chem. Ind.*, 1927, 46: 445–6.
30. See his 'Ten Years' Progress', *op. cit.* (3), 134, and his International Congress paper, 'State of the Cyanide Industry', cited in *J. Soc. Chem. Ind.*, 1903, 22: 766–7. Productive capacity was well above these demand levels. In this respect it should be noted that cyanide was used in a very dilute solution to extract gold – less than 1 per cent.
31. *CBM*, 8 July 1901, 5 August 1901 and 12 January 1904.
32. R.W. Moore, 'Commercial Cyanide of Potassium', *EMJ*, 1902, 73: 352.
33. See Tariff Hearings, *House Documents*, 62nd Congress, vol. 127 (Washington, DC, 1913), Schedule A (henceforth Tariff Hearings, 1913), 372–3.
34. See Browne, *op. cit.* (24), 186.
35. *The Mining Industry*, 1914, xi: 345.
36. See *The Mineral Industry*, 1920, 29: 312–1.
37. See Bueb, *op. cit.* (23), 485–6; Anon., 'The Manufacture of Cyanides', *EMJ*, 1902, 73: 728.
38. See D.W.F. Hardie and J. Davidson Pratt, *A History of the Modern Chemical Industry* (London, 1966), 62.
39. *CBM*, 30 October 1906.
40. *Hearings before the Committee of Finance, United States Senate, on the Proposed Tariff Act of 1921* (Washington, DC, 1922), *Vol. VI*, 4796 and 4800. The total consumption figure is approximate.
41. See *The Mineral Industry*, 1920, 29: 312–13.
42. *EMJ*, 1916, 111: 336. Note, however, that 97 per cent NaCN would not have produced results as good as pure NaCN!
43. *The Mineral Industry*, 1922, 31: 342.
44. *The Mineral Industry*, 1924, 33: 330.
45. See *The Mining Magazine*, 1924, xxxi: 191, 255; 1925, xxxii: 126; and 1925 (second half), xxxiii: 191.
46. See Document No. ICHO/HIS/0301, paper by J.P. Quinn, Box 308, ICI Archives, London. In the 1930s, however, the plant was shut down on economic grounds.
47. *The Mining Magazine*, 1925, xxxii: 126.
48. *Ibid.*, 116–20.
49. The latter, in Czechoslovakia, was from the Kaliwerke Aktiengesellschaft, which had been set up in Bohemia (Austria) just before World War I and given a quota, exporting mainly to South Africa. See Document No. CCCL/03C/0009, Box 66, ICI Archives, London.
50. See Document No. ICHO/CFD/3662, Box 426, ICI Archives, London. In 1930, DEGUSSA confined its own sales to Continental industry.
51. See Document No. 03C/0026, October 1931, ICI Archives, London.
52. A conclusion supporting that of H. Berghoff and R. Moller, 'Tired Pioneers and Dynamic Newcomers? A Comparative Essay on English and German Entrepreneurial History, 1870–1914', *Economic History Review*, 1994, XLVII: 262–87.

Putting the Wind up the Pilot: Cloud Flying with Early Aircraft Instruments

JOHN K. BRADLEY

Flying in Britain developed during the six years before World War I and was carried out invariably in good weather and clear visibility. The discretion to fit and use any of the few aircraft instruments available was left to the pilot, who had usually been taught to fly without instruments, as their use was thought to handicap his natural instincts. Pilots readily accepted the engine revolutions indicator, as it was essential to know their aircraft engine was developing maximum power to stay airborne. The airspeed indicator was next in importance, as the difference between the low airspeed at which the aircraft stalled and the higher speed at which it could suffer structural damage was only a few miles per hour. The altimeter was initially an instrument of achievement, recording the maximum height the pilot had managed to attain, but interest in it increased as rate of climb became important to military pilots.

Flying became more disciplined under war conditions. All service aircraft were fitted with revolution indicator, airspeed indicator, altimeter and compass, and pilots were taught the operation and use of the instruments, although they still learned to fly without them. Many pilots regarded instruments as useful only in emergencies, as when trying to return to the airfield following engine failure after take-off, or finding their way home after a dogfight, when lack of familiarity and use resulted in confusion and error. However, service flying also brought the need to operate in all weather conditions and to fly through cloud. In poor visibility or cloud, when the horizon was no longer visible, the pilot had no choice but to use his instruments, but it quickly became obvious that those he had were useless for the purpose.

Passengers on modern airliners, which take off in poor visibility and climb rapidly through cloud to emerge minutes later in bright sunlight, gain little insight of the difficulties faced by the World War I pilot in his open cockpit biplane without any form of artificial horizon or turn indicator. During the war, the compass remained the only cloud flying aid for service pilots. No gyroscopic turn indicators saw service in military or naval flying during the war, for although marine gyroscope technology was well advanced at this time, no small reliable gyroscopes existed that could be installed on aeroplanes. Fortunately, contemporary accounts by some pilots provide an idea of the difficulties they faced. One pilot described his experience of losing orientation in a cloud and emerging to find himself flying towards a vertical green wall. Only when he distinguished trees and a church sticking out of the wall did he realize what was happening.[1] Another very experienced pilot, Captain B. C. Hucks, gave a more graphic account three years into the war.

> On reaching 1,200 feet we got into a dense rain cloud, but carried on to beyond 5,000 feet, still in cloud, when the compass apparently began to swing (really it's the machine that begins to swing, not the compass), and efforts to check the compass had the effect of causing it to swing more violently in the other direction. The air speed then rushed up far beyond normal flying speed; all efforts to pull her up checked her only slightly; then the rudder was tried, back went the air speed to zero; there was an uncanny feeling of being detached from the machine, and I knew her to be literally tumbling out of the clouds. All efforts to settle down to a straight flight seemed to be unavailing, until we emerged from the cloud very nearly upside down. Assuming control again was then a easy matter.
>
> This sort of thing happened to me more than once, and in the Flying Corps vernacular 'puts the wind up you'. It has also happened many times to other pilots. In some cases they emerge from the clouds in a spin, in other cases the planes have collapsed under the strain of a sudden pull-up from a vertical nose dive. A few days ago, a squadron commander told me that on one occasion everything loose in his machine fell out whilst in the cloud.
>
> Flying in clouds is a very unpleasant state of affairs. First of all in cloud you see nothing whatever but your machine around you. There is no fixed point visible. The only means by which you can tell if you are flying in a straight course is by your compass and your air speed. The compass should give you your direction horizontally, your air speed your direction vertically. The first thing that happens, and very readily too if it's windy and bumpy, is that your compass card will begin to move slightly. It really appears to you that your compass has been affected by the cloud, and that you are still flying straight ahead. How often you hear a pilot say that when he got into cloud his compass started spinning. The moment the compass starts moving, it requires extremely delicate ruddering to get it back to a steady position; in fact one invariably over-corrects the compass movement and so the trouble begins. Once the compass starts on a good swing, I have found it nearly impossible to get it steady again until out of the cloud. Before your compass starts to move, the machine has already started to turn. You rudder the opposite way to check it, over correct it, and turn sharper the other way on a bank turn; then the nose drops and the speed goes up. Pulling back on your elevator lever has little or no effect, for if you are banked above

> an angle of 45 degrees the elevator has become a rudder. All this occurs without the pilot being in the least bit aware of the position that his machine is taking relative to the ground.
>
> The instruments available are of little use once he loses his control. It may be argued that if a stable machine is left alone, it will right itself eventually and assume a normal glide. It very likely would if the pilot could steel himself to let it entirely alone, but before he did so it would have to be left to do a sheer vertical nose-dive for a few moments, and in these days of big weights and little head resistance one is liable to attempt to pull-out too suddenly from the dangerously high speed attained in this dive.
>
> What I want to see fitted is an instrument that will show a constant vertical or horizontal line and be independent of centrifugal force. I have no ideas on the subject or suggestions as to how this is to be brought about, unless something in the nature of a small gyroscope driven by the air-screw could be employed in some way, but until something is provided I think we shall continue to have accidents from this cause.[2]

This kind of experience showed the urgent need to develop a turn indicator. The fact that Hucks was deliberately flying through cloud by 'ruddering' his machine to make a flat turn without banking shows he had been trained to navigate through cloud with an RAF Mark II compass.[3]

PROBLEMS OF VIBRATION AND ACCELERATION

Before we follow the development of the instrument Hucks wanted to see for cloud flying, it is necessary to be aware of the forces to which any aircraft instrument is subjected and the solutions that early designers adopted. There was no transfer of technology available from balloons, as balloons had no engines and, as they travelled in an air current, air speed measurement had no meaning. Instruments were used on airships, but these could be located away from the engines, and their size and weight were not too important. Horace Darwin, chairman of the Cambridge Scientific Instrument Company, described the difficulty of using any instrument on an aeroplane in the first Wilbur Wright Memorial Lecture, 'Scientific Instruments: their design and use in aeronautics', given in 1913. The subject of the lecture was both a recognition of the growing interest in instruments and an acknowledgement of Horace Darwin's authority in the field, as he had been a member of the government-appointed Aeronautical Advisory Committee since its foundation in 1909.[4] Darwin explained that;

> The aeroplane shakes, it does not remain level, and is subject to acceleration in all directions. The instrument should be designed to be unaffected by any of these disturbances. A vertical acceleration has the same effect as an increase in the downward pull due to

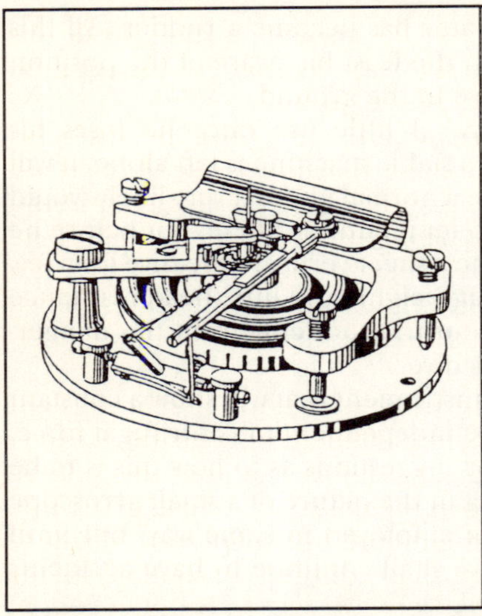

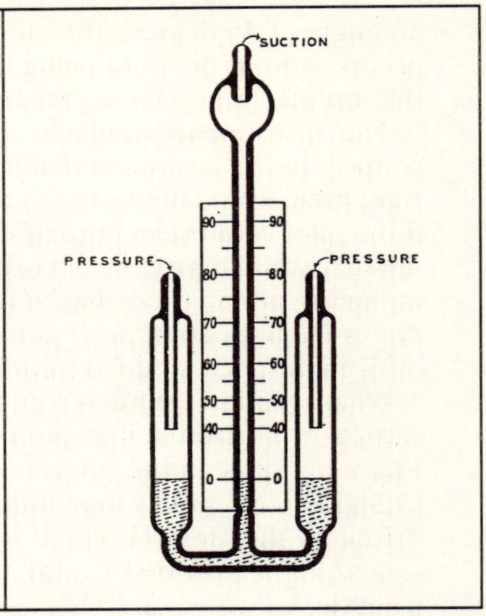

Figure 1 Altimeter design with rigid bridge mounting and strong spring to withstand vibration.
Figure 2 Short's 1912 velometer (Patent 14,749). Three limbs ensured that liquid in the centre limb was unaffected when tilted, and traps allowed the instrument to be inverted without liquid loss.

> gravity. A lateral acceleration has the effect of both altering the direction and amount of gravity. But vibration is the greater difficulty. The hand of an instrument may move so much and so rapidly that it is difficult to estimate the mean reading on the scale.[5]

He recommended that instrument makers adopt the geometric design practices of Clerk Maxwell, and that any dial type aircraft instrument would need to have a carefully balanced mechanism to give a steady reading. Instruments involving weaker measurements, like the airspeed indicator and compass, presented more design difficulties than the altimeter and revolutions indicator, which involved stronger measurements. The aneroid barometer movement (Figure 1) had already been converted to an acceptable altimeter by replacing the two-point bridge mounting with a more rigid three-point mounting by Elliott Brothers, Casella, Munro and Negretti and Zambra.

Revolution indicators, using spring constrained rotating weights, had been available for four years from Elliott Brothers and S. Smith & Sons, who had quickly adapted their motor vehicle instruments to the aeroplane. As early as 1910, Elliott Brothers received praise for their indicator from T. O. M. Sopwith after his record breaking long distance flight to Belgium,

> I should like you to know how essential your revolution indicator has

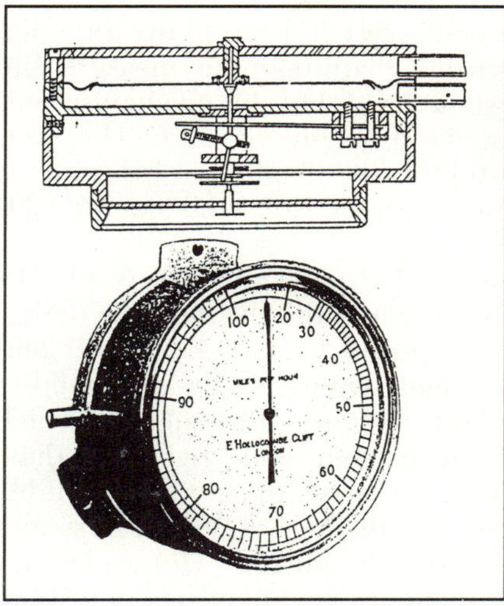

Figure 3 Clift airspeed indicator of 1913.

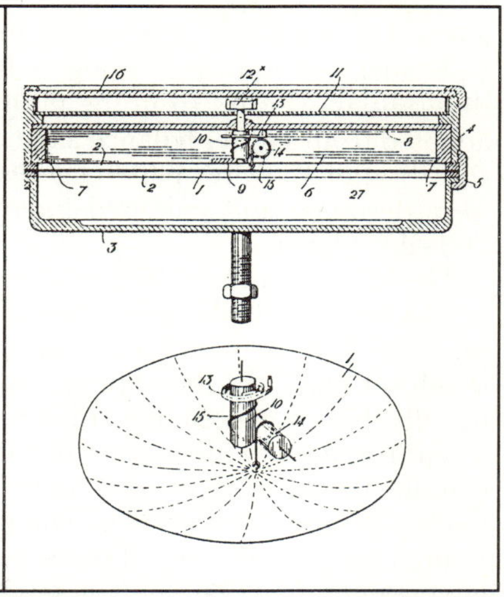

Figure 4 Ogilvie airspeed indicator of 1913, showing thread and roller arrangement.

been to my flying. I would not be without it for anything. I practically 'drive on it' the whole time, as well as do my engine tests with it. During my flight to Belgium, it worked magnificently, never giving a moment's trouble and showing a perfectly steady reading.[6]

Sopwith did not write a similar letter to his compass maker as his compass card continually rotated during the whole flight and was the reason he ended in Belgium instead of France. Although mechanical revolution indicators gave a steady reading, their serviceable life was relatively short as the engine vibration on aeroplanes was more severe than that on motor vehicles and caused high rates of wear to instrument bearings and gear drives. Nevertheless, they remained in use for 15–20 years.

Air speed was initially measured with a rotating cup anemometer, a pressure plate or simply a piece of string, which had to be mounted some distance away from the pilot to be outside the slipstream from the propeller. Makers included J. J. Hicks and Negretti and Zambra. The first reliable air speed indicator was a U-tube manometer called the velometer (Figure 2), designed and patented by Frank Short at the Royal Aircraft Factory at Farnborough in 1912. Connection to a pitot-static head on a wing strut allowed the measuring element to be well away from the slipstream, while the indicator could be placed in front of the pilot. The velometer was manufactured by Casella and Elliott Brothers and was often supplied to military aircraft on a standard instrument panel. Darwin drew attention to the velometer in his paper as the only airspeed indicator that could withstand vibration at the time. However, being 'gravity controlled', its reading was affected by acceleration forces and it was not a convenient instrument to read. He had tried to design a dial type airspeed indicator using a flexible diaphragm to measure the differential pressure from the pitot-static head, but the mechanism was adapted from a letter balance and could not tolerate vibration. However, shortly after Darwin's paper, successful flexible diaphragm instruments were designed and patented independently by Eric Hollocombe Clift and Alec Ogilvie.

The Clift instrument (Figure 3) consisted of an air chamber divided by a light flexible diaphragm of specially prepared thin leather, the deflection of which actuated a pointer mechanism by a rocking lever and quadrant plate.[7] In the Ogilvie instrument (Figure 4), the air chamber was divided by a stretched rubber diaphragm, the movement of which was transmitted to the pointer by a thread passing round an intermediate spindle before making two ascending turns on the pointer spindle. Ogilvie designed this novel mechanism with the help of Booth Harris and Frank Cowlin – all three of them Cambridge graduates.[8] Horace Darwin later commented that when he saw any new instrument he had one of two thoughts, either 'Why didn't I think of that', or 'What fool expected that to work', and he would have put Ogilvie's crude but effective indicator mechanism into the second category, had he seen it before it was proved.

As the flexible diaphragm air speed indicator mechanism was spring controlled it was unaffected by acceleration forces. However, it did have a significant measurement lag, as the volume of the chamber had to be large to make it sensitive, and this lag was further aggravated by the hysteresis of the flexible diaphragm. The instrument depended on continuous use to keep the diaphragm flexible and the instrument in calibration. One technique practised when taking an indicator from stores was to connect it to one arm of a rubber U-tube, while the other arm was filled with water and placed under a running tap for several minutes, so that the pressure fluctuations flexed the diaphragm.

However, the convenience of reading the dial indicator for routine flying completely outweighed its disadvantages and once Clift and Ogilvie instruments had gone into quantity production, all service aeroplanes were fitted with them by 1916. They remained in use throughout the war, but eventually the freezing temperatures encountered as operational altitudes continually increased made the diaphragm inflexible, and instruments with metal diaphragms had to be developed.

EARLY AIRCRAFT COMPASSES

The aircraft compass was to prove the most difficult instrument to develop. Shortly after S. F. Cody designed the first aeroplane to fly at Farnborough, Captain Creagh-Osborne at the Admiralty Compass Department wrote to him in August 1909 suggesting that a Navy liquid-filled compass, which could withstand gunfire shock, might be suitable for an aeroplane.[9] He flew with Cody the following month and, finding that the compass card continually rotated owing to the engine vibrations, placed the compass in a box filled with cotton waste to reduce the effect, after first proving the idea on the floor of a car driven by a single cylinder engine. A few months later, he discussed compass requirements with the pioneer aviators at Brooklands, and wrote to the Glasgow compass makers Kelvin and James White outlining the requirements and encouraging them to be first in this new field. They produced their first aeroplane compass the following year, which was a smaller version of the Chetwynd naval type compass, weighing just over 3 lbs (1.35 kg) and housed in a circular case filled with horse hair. The arrangement for attaching the case to the airframe was novel. The case had an external circumferential groove in which a girdle of stranded wire was wound, from which three ligaments of stranded wire allowed it to be attached to any convenient part of the aeroplane, the wire also acting as a shock absorber.[10] The designers stressed the importance of mounting the compass away from ferrous objects or replacing them with phosphor bronze, drawing particular attention to the steering gear which was constantly changing position in flight.

E. H. Clift was critical that the Kelvin and White aircraft compass did not have gimbals to keep it level and used horse hair suspension to overcome shock and vibration, and by 1912 was advertising his own compass, which, he claimed, overcame these limitations.[11] He retained the reduced diameter and slightly buoyant card, but provided rigid cross-shaped rubber mounts to secure the outer gimbals to the frame, noting that if the period of the card agreed in any way with the rate of vibration of the aeroplane, the whole matter became hopeless. He stressed the importance of both correctly positioning and swinging the compass, as he found that aviators failed to understand that a small piece of iron close to the compass could have a much greater effect than the engine at some distance from it. He alerted them to the problems arising from brass control levers strengthened with steel tubes, from adjacent brass instruments fixed with steel screws and from energetic assistants who leaned over the cockpit with tools, bolts or spare parts in their overalls, while the whole effort could be wasted if the pilot himself later carried a steel map case or a bag of tools with him. He also introduced his leeway indicating compass, in which the bottom of the bowl was transparent. This allowed the aviator to observe the direction traversed by objects on the ground and correct his drift from a beam wind by altering the direction he was flying until the objects on the ground moved parallel to the desired course. Clift's compasses and navigational aids were made by Henry Hughes and Sons of 59 Fenchurch Street, London.[12] Other makers also advertised aircraft compasses, but it is doubtful whether any of their compasses would work in the air, as none had any form of anti-vibration mounting and they would cease to point north once the engine was started. Eventually, a conference was held in December 1912 to discuss the standard features required for service aeroplane compasses, and a specification was produced and given to manufacturers on the Admiralty

List. The resulting Pattern 200 compass became the standard military and naval compass for the next two years.

WARTIME COMPASSES DEVELOPMENT BY KEITH LUCAS AT FARNBOROUGH

The problems with early compasses continued as improved engine power increased the vibrational and accelerative forces they were subjected to. Even before World War I, service reports noted that Clift compasses were unsatisfactory and that although the Pattern 200 from Kelvin and White was preferred to their older Chetwynd type, it should be mounted in gimbals as it jammed in bumpy weather.[13] There were reports of compasses sticking in turns, of compass deviation in cloud and of balloon ascents to measure whether the earth's magnetic field varied with height, which might explain the deviation.[14] Regarding vibration, a naval pilot ironically commented that, 'Life is at its darkest hour when the compass shows signs of settling down and at last gives more or less a steady reading', because when this happened it meant that one's engine had failed.[15] The compass was obviously still in trouble, and as it was the only instrument available to assist service pilots to fly through cloud, it was serious trouble.

Keith Lucas, the Cambridge physiologist, joined Farnborough in September 1914, persuaded by Darwin that he could achieve more for the war effort there than by enlisting in the army. His interest was immediately drawn to compasses, when on his first day he was handed a report from the Royal Flying Corps in France complaining that its compasses were useless.[16] Undertaking a few simple ground and flight tests, he quickly confirmed that engine vibration could cause heading errors of between 40 and 50 degrees at full engine revs, and then proceeded to design a better compass. His investigations and the design basis he proposed for a new compass are fully documented, and included the theoretical investigations of the subject by Darwin, Greenhill and Mallock of the Aeronautical Research Committee, who approached the problem as a mathematical tripos question.[17] However, Darwin made an important suggestion, that vibration could be overcome by reversing the cup and pivot of the compass, so that the pivot was on the magnet and rested in an inverted cup in the bowl of the compass. He gave no explanation of why the inverted arrangement would reduce the error, though the analogy of a conjurer spinning a plate on a stick, a feat difficult if the stick was fixed to the centre of the plate, was later used. Reversing the coupling between two elements of a mechanism to provide better performance was a technique in the geometric design of instruments practised by Darwin. He proved that his inverted support reduced vibration error on a vibrating table, and showed also that vibration error was caused by friction when obtaining the same error by replacing the compass magnet with a brass rod.[18] Adopting his suggestion, Lucas found that when the cup and pivot on the Admiralty Pattern 200 compass were reversed, the heading error at full engine revs both on the ground and in flight could be eliminated. He also observed, as Clift had found earlier, that if the period of the compass was similar to that of the aeroplane, the problem persisted.

With a few simple experiments, Lucas had found and eradicated the compass vibration problems that had plagued the aircraft compass for 5 years, and he next turned his attention to the heading error caused by tilting the axis of the card. It was well known that if the axis of a compass card was gradually inclined towards the east or west, the vertical component of the earth's magnetic field produced a moment about the axis to pull the north point of the card downwards. Lucas realized that this well known phenomenon applied to an aircraft compass as the axis of a compass card tilted when the aeroplane accelerated or decelerated while flying straight and level, or when banking into a turn.

It was during turns that pilots complained that compasses were *sticking*. Many pilots also reported that their compasses went wrong in clouds, and they found themselves coming out of clouds on the same side they had entered. Those were the lucky ones, as most pilots who lost orientation in cloud got into a spin, which was usually fatal as the procedure for recovering from a spin was unknown at this time. In fact, it was unwise to fly any aircraft powered by a rotary engine into cloud, as the gyroscopic torque of the rapidly rotating cylinders invariably produced an uncontrollable spin once the pilot lost sight of the visible horizon.

Perhaps the best way to visualize these reports is to realize that the compass card must continue to point north throughout any turn if the true heading of the aircraft is to be known. If the pilot is flying due north and begins a turn towards the east, and the card happens to turn at exactly the same rate as the aeroplane, the pilot will be given the impression that the aeroplane is not turning or that the card has stuck. If, on the other hand, the compass card turns more rapidly than the aeroplane, the pilot will think that the turn is in the opposite direction to that which is actually occurring. This in cloud, with the horizon invisible, would lead the pilot to further increase his rate of turn eastward to correct the apparent turn towards the west and eventually result in a spin. Lucas coined the term 'northerly turning error' to indicate this compass behaviour. Initially, the Admiralty Compass Department refused to believe that any compass could behave in a manner of which they were not already aware, but were convinced after Farnborough got Creagh-Osborne to fly with them and observe the effect for himself.

Lucas concluded that the best solution was to reduce the magnetic moment of the compass by giving it a longer period that would be slow to respond to the vertical component of the earth's magnetic field. To reduce disturbance to the compass card by the movement of the liquid when the aircraft pitched and rolled, his new compass (Figure 5) had a spherical bowl and a narrow cylindrical card, which used Darwin's inverted cup and pivot and had plenty of clearance between the card and the inner surface of the bowl to reduce skin effect friction. Anti-vibration mountings were provided by four leaf springs, which retained the spherical bowl in a square housing. The final design became the RAF Pilot's

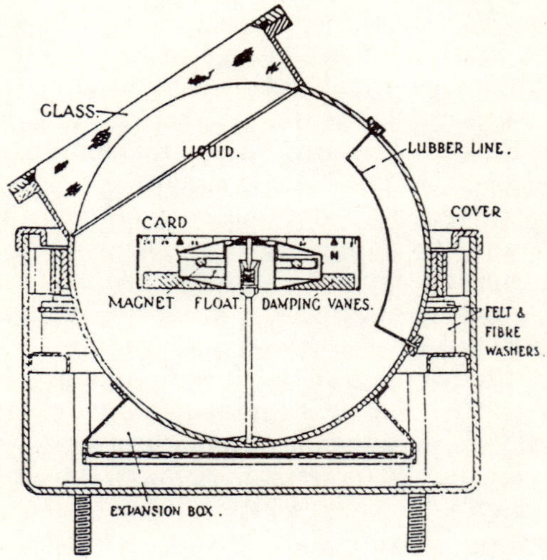

Figure 5 Lucas RAF Mark II long period compass.

Compass Mark II. To reduce the vertical component of the earth's magnetic field further, the pilot was taught to steer the aeroplane in cloud without banking, by using only the rudder and not the ailerons. With care, a trained pilot could navigate through cloud, but if he lost control he could not regain it within the cloud, and had to hope there was time and space to recover when he emerged below it and could again see the visible horizon.

Tragically, Keith Lucas was killed on 16 October 1916 in a mid-air collision while learning to fly at Upavon, but his work on the compass was continued by F. A. Lindemann (later Lord Cherwell), who had himself learned to fly on the same Royal Flying Corps (RFC) course as Lucas. In 6 months, Lucas used a scientific approach to produce a new compass, which was used by the RFC throughout the war. There was nothing remarkable to him in the way he examined compass behaviour: it was just a problem requiring an instrument solution, no different from his pre-war physiology experiments. His direct approach to experiment illustrated that an experienced research scientist could quickly analyse a problem and develop a solution in an unfamiliar field, and undoubtedly went some way towards the scientific method that Haldane had envisaged in 1909. However, his experiments were still based on short tables of discrete flight test observations, and it was not until 1919 that the complex behaviour of the aircraft compass was reproduced in a laboratory by H. E. Wimperis in the Air Ministry Laboratory at Imperial College.

Because the long period compass assisted cloud flying, Farnborough felt that it had met the immediate needs of all service pilots, and did not consider that most pilots simply required a responsive compass for routine navigation in good visibility. This included Royal Naval Air Service (RNAS) flying boat pilots, who routinely flew long duration submarine patrols in poor visibility, but did so at wave-top height, which provided them with a horizon reference plane. The Admiralty Compass Department, which was never happy with the idea of a compass with a reduced magnetic moment, produced its own range of quick period compasses for normal flying. Nevertheless, what was steadily becoming obvious to everyone was that some form of turn indicator or artificial horizon device was required for cloud flying, and this would probably involve the use of a gyroscope once a small reliable one could be developed. This did not materialize until 1919, and in the interim the static turn indicator, which measured the difference between the hydrostatic pressure at the wing tips of a banked aeroplane, provided a valuable alternative.

DARWIN'S STATIC TURN INDICATOR

The development of the static turn indicator was described in a paper by Horace Darwin in 1919.[19] Work had started in 1916 when Alec Ogilvie, at the Air Board, discussed with Henry Tizard, at the Aeroplane Testing Squadron at Martlesham Heath, the difficulties experienced in cloud flying, in fact in flying any straight course at all. They considered that a pitot head to measure the difference in velocity at each wing tip might be the basis of an idea to fly a straight line. However, their experiments proved unsatisfactory and Darwin, now Chairman of the Air Inventions Committee, offered to take the whole job over. Calculations at the Cambridge Scientific Instrument Company quickly showed that the differential pressure between the pitot tubes at the wing tips during a banked turn was counterbalanced by the difference in the centrifugal force in the air tubes taking the pressure measurement to the cockpit. However, this small centrifugal differential pressure (which mathematically is equivalent to the difference in the hydrostatic pressure between the wing tips in a turn) could be used to indicate bank providing it could be measured (Figure 6). Darwin attributed eventual success to the experimental work at Farnborough and Orfordness to improve the sensitivity of the Ogilvie pressure gauge. F. A. Lindemann, who flight-tested the new device, demonstrated for the first time that it was possible safely to fly and manoeuvre an aeroplane in cloud.

In the discussion following Darwin's paper, H. E. Wimperis said that recent navigational trials in Newfoundland, in preparation for a trans-Atlantic flight, had found that the accuracy of sextant observations was materially increased when the aeroplane was flown level with the help of a static turn indicator. He was able to compare the performance of the static turn indicator and a gyroscopic turn indicator currently being tested, having recently flown in a Handley Page bomber in which both

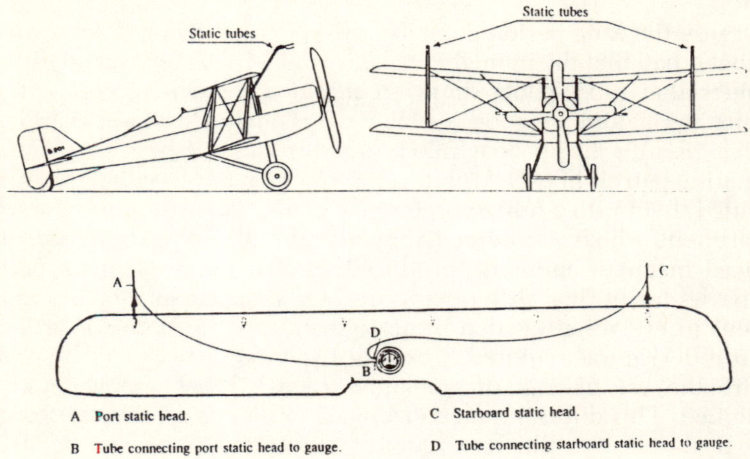

Figure 6 Darwin's static turn indicator showing location of static tubes on wing tips and connection to sensitive Ogilvie type pressure gauge.

instruments were installed: 'It was almost uncanny how they both did the same thing, the only difference being that the Darwin type had a small lag due to the compressibility of the air in the tubes.'

Its ability to measure a hydrostatic head of a few millimetres between the wing tips of an aeroplane travelling at over 100 m.p.h. made the static turn indicator a remarkable instrument. Although it was successful, its development came too late to influence wartime flying and only 70 were delivered to the Western Front before the armistice in 1918. However, its real value was to separate the development of the turn indicator from the development of the compass, allowing the compass to revert to its proper role as a navigational instrument.

PENDULUMS, FRENCH TOPS AND GYROSCOPIC TURN INDICATORS

While the British pursued the Lucas long period compass and the Darwin static turn indicator, the French were developing spinning top gyros. These were unconstrained gyros, similar to a child's spinning top, in which the gyro axis moved under a graduated glass dome to indicate the turn. The RFC tested those by the Garnier and Hebrard companies, but found they were insensitive to gradual turns, ceased to function above 45 degrees when the gyro struck the case, and were generally much coarser than the static turn indicator.[20] As they did not indicate gradual turns they were unsuitable for cloud flying and were never adopted into British service, although they remained popular with French pilots throughout the war. Testing the French spinning top gyros was useful to British research as it showed that this form of gyroscope was not the way to go. In the USA at this time there was no research on aircraft turn indicators and

Sperry's first turn indicator patents were based on the British static turn indicator and the French spinning top gyro.[21]

There was no shortage of patents during the war purporting to indicate the true horizontal or true vertical in an aeroplane. Most were based on the pendulum, as the inventors did not appreciate the acceleration forces to which a pendulum was subjected when mounted on an aeroplane, or, if they did, they thought that enclosing it in liquid would reduce the forces. The inventions usually adopted some gravity-sensitive element, such as a weighted plummet or bubble enclosed in a liquid filled sphere, hemisphere or annular ring. Patents for impractical aircraft instruments became so numerous that Darwin, as Chairman of the Air Inventions Committee, published a notice concerning the hundreds of inventions and suggestions for aircraft instruments that had been received which disregarded the action of centrifugal force. He stressed that, 'although it was possible that improved forms of instruments may be produced, it is scarcely likely that this can be done by anyone without the necessary scientific and mechanical knowledge required for an investigation of this nature'.[22]

The gyroscopic rate of turn indicator was patented by Professor Henderson at Greenwich Naval College in 1908, but the patent was secret and the instrument was subsequently reinvented by H. E. Wimperis in 1910, and again in later years by Sperry, Mendendall and Lanchester. However, Henderson did not assemble a prototype of his turn indicator for testing until 1918, and only then because he could not understand why the static turn indicator was being used, unless it was for weight reasons. His device was a constrained gyro, belt driven from the engine via a shaft and universal joint, and weighed some 20–24 lbs (9–11 kg). The prototype performed successfully at Martlesham Heath in August, and was then sent to Farnborough to be redesigned as a more practical instrument. The redesign was completed in November 1918 by F. A. Lindemann and W. S. Smith, who drastically reduced the weight to under 5 lbs (2.25 kg) by designing a new instrument in which the gyro rotor was placed outside the fuselage and driven by the slip stream. Lindemann's pilot's log shows that he made three flights in December 1918 to flight test this instrument before going to Oxford as Professor of Experimental Philosophy in January 1919. The new turn indicator quickly entered service as the RAE gyroscopic turning indicator (Figure 7), but service conditions exposed a bearing failure problem – the slipstream that provided ample thrust to drive the rotor also drove dust into the bearings. Dust caps were provided and a brake for the rotor when not in use, but both modifications failed to overcome the excessive bearing wear and after exhaustive tests it was finally abandoned in February 1923, by which time an electrically driven gyroscopic turn indicator was available, which was contained within the fuselage and called the Kingsway after the address of the RAF research and development department.[23] Nevertheless, the exposed rotor turn indicator, like the static turn indicator that preceded it, had served its purpose as an intermediate step towards a more effective instrument for cloud flying.

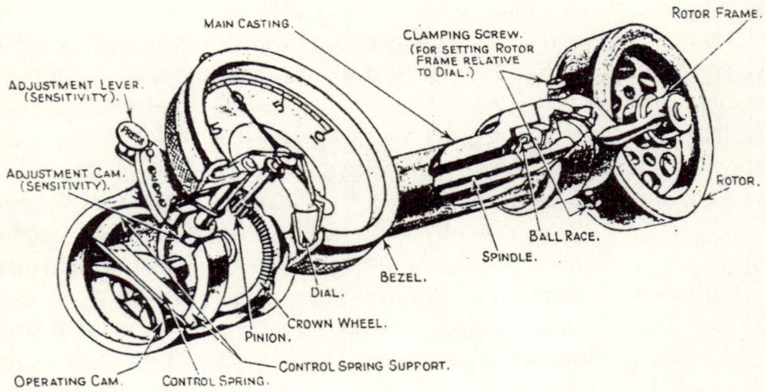

Figure 7 RAE turning indicator designed by Lindemann and Smith (Patent 147271/19). The gyro was spring controlled and designed so that the rotor was outside the fuselage, directly driven by the slipstream, and the indicator was inside the fuselage in front of the pilot. The web of the rotor was drilled with a series of oblique holes to produce a high driving torque. The axis of the rotor was parallel to the fore and aft axis of the aeroplane and mounted in a fork secured to a rotatable spindle. A turn to the right with this configuration caused the fork and spindle to rotate in a clockwise direction and the movement was transmitted to the pointer by gearing. Similarly, a turn to the left produced an anti-clockwise rotation of the spindle.

SUMMARY

There are few papers and books to date on the history of aircraft instruments, for a period which now spans some 80 years.[24] The aeroplane was different from previous forms of transport, because it did not come safely to rest when its engine or part of its structure failed. Instruments were introduced to reduce the likelihood of this occurring, by indicating the condition of the engine and the narrow speed limits between stall and disintegration. Suitable engine revolution indicators, air speed indicators and altimeters were available when World War I began, but war brought new requirements, particularly the need to fly in cloud or poor visibility, and the compass and turn indicator were developed to assist the pilot. The rate of climb indicator, measuring the pressure in a vacuum flask with a capillary leak to atmosphere, was introduced for test flying but was not used in service aircraft.

Any instrument fitted to the airframe was subject to severe vibrational and accelerative forces that were not present on the ground, and the instrument designer was faced with the difficulty of designing sensitive mechanisms that could tolerate these forces. On instruments like the aircraft compass, anti-vibration features became essential if such instruments were to work. The cramped open cockpit and small instrument panel severely limited both the size of instruments and the freedom to locate them for the convenience of the pilot, while low engine power made weight reduction the priority for all airborne equipment. The accelerative forces were difficult enough for scientists to appreciate, but were often incomprehensible to service pilots, and led to their mistrust of aircraft instruments in general, and the compass in particular.

Early aircraft instruments were mechanical, self-contained, self-powered devices. Air velocity and pressure measurement had been established in meteorology and mountaineering, and the compass was well understood in marine navigation, but the transfer of this technology to the aeroplane and its environment required considerable resources and experimentation for a relatively small number of instruments. The problems to be overcome were not new, but the combination of them was, and what appeared scientifically easy in theory proved technologically difficult to achieve quickly. The manometer proved to be the most adaptable early aircraft instrument in the absence of better technology.

Direct association with flying became more important than established instrument making practice in the design of new aircraft instruments. Rapid wartime development was due to the foresight of R. B. Haldane, who believed that aeronautical research should be on a scientific basis, and to Horace Darwin, who encouraged many able university research scientists into military aviation. With little previous experience of the subject or the environment, these scientists used standard laboratory research practice to interpret quickly apparently inexplicable instrument behaviour, like that of the magnetic compass in an aeroplane. The university scientist had merely changed his lab coat for a flying suit. In the period examined, there was little in the way of equipment like environmental chambers to observe the behaviour of aircraft instruments in the laboratory. The aeroplane itself remained the only place to test instruments that were needed for it. The wartime use of civilian scientists in aircraft instrumentation was not an isolated example; sound ranging and submarine detection required the same kind of applied scientific effort and even simple devices like the trench periscope and daylight signalling lamp took more effort than would be imagined.

The development of the static and gyroscopic turn indicator was a British one, funded entirely by the government, and was undertaken by the RFC with no RNAS involvement, as they did not share the same interest in cloud flying. F. A. Lindemann played an important role in the whole development, test flying and improving the long period compass designed by Keith Lucas, and later the static turn and the gyroscopic turn indicator, a two-year wartime effort ignored by his biographers in favour of his more famous work on spinning, which took little more than two weeks. The gyroscopic turn indicator was a fast spinning, pivoted mechanism, requiring power to drive it, which was significantly different from previous aircraft instruments. Military cloud flying prepared the way for commercial cloud flying, where the question was no longer one of *safely* flying through cloud, but of passenger *safety* in doing so, a subtle difference between military and commercial flying. Nevertheless, the instrument at the beginning of this paper that Hucks hoped to see on an aeroplane had become a practical reality.

Notes and References

I wish to acknowledge the encouragement and suggestions from Graham Hollister-Short for writing this paper, which is taken from research for a PhD thesis, The History and Development of Aircraft Instruments – 1909 to 1919', Imperial College, London, 1995.

1. H.R. Brooke-Popham, 'The Air Force', *Jnl of United Services Institute*, 1920, **65**: 43–70.

2. B.C. Hucks, 'A Further Three Years of Flying Experience', *Aeronautical Journal*, 1917, **21**: 426–9, read 6 June 1917. D. Young, *Rutland of Jutland* (London, 1963), p. 18, explains that the origin of 'put the wind up you' came from flying in open framework aeroplanes, in which if you stalled, instead of the air rushing towards you, it came rushing upwards, literally 'putting the wind up you'.

3. RAF at this time denoted the Royal Aircraft Factory at Farnborough, which did not become the Royal Aircraft Establishment (RAE) until the Royal Flying Corps (RFC) became the Royal Air Force (RAF) in 1918.

4. The Secretary of State R.B. Haldane set up the Advisory Committee for Aeronautics, as in his view the problems of flight were too complex for non-scientific enthusiasts. He wanted to secure for the government the services of qualified scientists, who could discover the principles of flight to enable the development of efficient military aircraft. He saw that aircraft development required similar experimental techniques to ship design, where hydro-dynamic problems were worked out with models in an experimental tank before the ship was built. Haldane asked the chairman of the National Physical Laboratory Lord Rayleigh, and its director R.T. Glazebrook, to prepare a scheme that would secure the co-operation of the NPL with the embryo aeronautical services, which included the Army Balloon Factory at Farnborough. Rayleigh and Glazebrook appointed several Cambridge colleagues to the new committee, i.e. H. Darwin, W.N. Shaw and R.H. Mallock.

5. H. Darwin, 'Scientific Instruments: Their Design and Use in Aeronautics', *Aeronautical Journal*, 1913, 170–90.

6. T.O.M. Sopwith, 'Correspondence', *Flight*, 21 January 1911: 48.

7. The background to the design is somewhat unclear. Clift lived in Earls Court and had previously patented some aircraft compasses and navigational aids. He approached the instrument maker J.W. Munro to help design a manometer type airspeed indicator, but Munro's chief designer Rooker produced instead a flexible diaphragm design, which J.W. Munro patented (Patent No. 1518/13). However, they appear to have assigned the patent back to Clift under some business arrangement, as instruments were afterwards ordered through Clift and manufactured by Munro on a royalty basis until 1917, when S.A. Smith & Sons bought out Clift and took over manufacture.

8. It was patented (Patent No. 13796/13) in the name of the British Wright Company, of which Ogilvie was a director. This company had been formed in 1913 by the Chancery Lane patent agent Griffith Brewer with Ogilvie and others to protect the Wright Patents in the UK. Initially, Elliott Brothers made the instruments, but during World War I Griffith Brewer established the Aeroplane Instrument Company at 9 Galen Place, London WC1, to manufacture them.

9. National Maritime Museum, Ref 43–521, Lt G.H. Alexander (Chief Compass Examiner 1907–30), *Compass Catalogue and Explanatory Notes*, with additions by Cmdr W.E. May (Naval Assistant 1928–30), 231–9. Also A.E. Fanning, *Steady as She Goes: The Official History of the Admiralty Compass Observatory* (HMSO, 1986).

10. M.B. Field and F.A. King, 'The Evolution of the Aero Compass', *Flight*, 5 October 1912: 891–2.

11. E.H. Clift, 'Correspondence', *Flight*, 13 May 1911: 429; 'The Magnetic Compass – Its Construction and Use, *Flight*, 24 February 1912: 163–5, and 2 March 1912: 199–201; and 'A New Compass', *Flight*, 20 July 1912: 661.

12. Editorial, *Flight*, 28 March 1914: 347. It is possible that Hughes designed as well as manufactured the compasses for Clift, as Munro did for his air speed indicator, but there is no documentary evidence to support this. Hughes certainly had considerable experience

with aircraft compasses when they assisted Creagh-Osborne with his range of wartime ones.

13. PRO/AIR 1/777/204/4/435, 'Supply and performance of compasses', correspondence between Squadrons and RAF, Farnborough, 1913–14.

14. PRO/DSIR 23/321/f and DSIR 23/366 W.R. Crocker (Lt), Various reports, undated.

15. Members of Squadron, *Naval Eight: The History of No. 8 Squadron RNAS* (London, 1931).

16. R.F. Mayo in Alys Keith-Lucas (ed.), *Keith Lucas* (Cambridge, 1934).

17. K. Lucas, *The Error of Compasses on Aeroplanes, Part 3 of R&M 238*, and *On a New Type of Compass for Use on Aeroplanes, Part 4 of R&M 238* (HMSO, April 1915).

18. PRO/DSIR 23/321/G, 'Report on Compass Experiments', Horace Darwin, 24 February 1914.

19. H. Darwin, 'The Static Turn Indicator for Aeroplanes', *Aeronautical Journal*, 1919, 23: 617–25.

20. PRO/AIR 1/1199/204/5/2607, Orfordness Report D-83, Gyroscopic Indicators, 1918.

21. Sperry patented a static turn indicator (Patent No. 133,714, October 1918) using a venturi at each wing tip, which Farnborough had found unsatisfactory. His first gyroscopic patent (US Patent No. 1,309,489, 8 July 1919) adopted the French spinning top gyro and overcame wander by providing two tops spinning in opposite directions, leaving it to the pilot to decide which one to use! His gyroscopic turn indicator (US Patent No. 1,443,102, 24 October 1922) was gravity controlled, with springs to prevent it precessing too far, and according to T.P. Hughes, *Elmer Sperry* (Baltimore, 1971), 235–6, suffered problems of acceleration and of being air driven.

22. The notice appeared in numerous journals, e.g. H. Darwin, 'Aeronautical Inventions', *Aeronautical Journal*, 1918, 22: 207–8.

23. PRO/DSIR 23/9524, 'Present Position on Turn Indicators, Report by Sqdn Ldr R.M. Hill, 2 May 1924.

24. What history is written examines the need for the instruments and the arrangement of their display to the pilot, e.g. L.F.C. Coombs, *The Aircraft Cockpit* (Wellingborough, 1990), and where the design of early instruments is mentioned, it is with amusement rather than with any concern for achievement, e.g. R.A. Chorley, 'Seventy Years of Flight Instruments and Displays', *Aeronautical Journal*, 1976, 70: 323–42. Textbooks on aircraft instruments appear from 1920, but although they describe the principles and construction of instruments, they do not explain why these principles and construction were adopted, e.g. C.J. Stewart, *Aircraft Instruments* (London, 1930).

'Seeing' Is Believing: The Development of Microwave Radar in Britain, Summer 1940

STEPHEN N. TRAVIS

INTRODUCTION

Microwave radar was developed in Britain in response to a specific threat, that of the German night bomber. The reason for this perceived need can be traced back to the decision to use scientific means to investigate how to protect Britain from air attack, which was taken in 1934. In early 1935, the Committee for the Scientific Survey of Air Defence (CSSAD), under Sir Henry Tizard (then rector of Imperial College) approached Robert Watson-Watt of the National Physical Laboratory. Watson-Watt had experience of ionospheric research using radio waves, and the CSSAD wanted to check if radio waves could be used as a means of detection. Successful trials, showing that this principle could indeed be used, led to the development of a research programme based at Bawdsey Manor in Suffolk, and the eventual installation of an early-warning radar chain, known as Chain Home, around the south and east coasts. This radar chain, in conjunction with a fighter-control network to direct RAF fighter aircraft, was used to great effect during the Battle of Britain in 1940.

In 1936, the far-sighted Tizard realized that if the radar chain was successful, enemy bombers would be forced to operate at night. The radar chain, and operators, were only able to place fighters within five miles of incoming bombers. While this was perfectly sufficient in conditions of good visibility, it would not be so when conditions were foggy, or at night-time. Tizard suggested that work should commence on designing an airborne radar. This task was handed to Dr E. G. Bowen, who, by 1939, had managed to construct and test an airborne radar working on a wavelength of 1.5 m. This set, developed with the help of only a tiny team, was a remarkable achievement given the contemporary low-power valves. Unfortunately, 1.5 m airborne interception (AI) radar suffered from a fundamental flaw.

There is a physical law which links the narrowness of a beam of radio waves to the dimensions of the antenna from which it is transmitted. Owing to the limitations of the size of antenna that could be fitted in an aeroplane, the beam of the 1.5 m AI was very wide, operating on a 'floodlighting' principle. This fact limited the range at which the set could be used to detect aircraft to its height above ground. Put another way, enemy aircraft could reduce the risk of detection by flying lower.

Bowen recognized this flaw, and he believed that the solution to the problem was to construct a set that operated on a wavelength of around 10 cm. This would produce a very narrow, searchlight-like beam from the size of antenna which he believed could reasonably be fitted into an aircraft. Soon after the outbreak of the war in September 1939, he discussed this problem with Sir Charles Wright, the Director of Scientific Research at the Admiralty. Wright believed that the Navy needed a similar narrow-beamed radar for gun direction. As no high-powered transmitter valve that operated on such a wavelength existed at this time, he used his position on the inter-service Committee for Valve Development to commission research into such a valve. This culminated in Randall and Boot's cavity magnetron, developed at Birmingham University and refined at GEC's Wembley laboratory during the winter and spring of 1940.[1]

At the outbreak of war the whole of the British Air Ministry Research Establishment (AMRE), as the radar research centre then became known, was evacuated from Bawdsey Manor, which was thought too vulnerable to the air attack that was immediately anticipated. Their destination was Dundee University, where they were not expected and there was no room. To make matters worse, Bowen's airborne group was moved to Scone airport, some 20 miles away. To add insult to injury, the group was moved again in November 1939 to St Athan in South Wales.

At the outbreak of war, many university scientists were recruited to work for the war effort in government research. Many of these were to become eminent in their fields after the war, but at first they sat uneasily within the civil service structure of government science. One such university scientist was Dr (later Professor Sir) Bernard Lovell, who was working on cosmic ray research at Manchester University. He was recruited because his professor, Patrick Blackett, was a member of the CSSAD. Lovell was set to work on installing the 1.5 m prototype AI into service aircraft, along with the other members of the airborne group to which he found himself attached. Their move to St Athan, where they worked in unheated hangars during the worst winter for fifty years, left little time for him to pursue the research into 10 cm antennas which Bowen had allocated to him. However, the realization in higher circles that effort was being wasted in this way led to an improvement in conditions, so that Lovell was joined by Dr (later Sir) Alan Hodgkin in February 1940. The two commenced research into antennas at this time.[2]

At the same time, this realization created the impetus for re-establishing AMRE at a single location after its split between Scotland

and South Wales. The location chosen was Worth Matravers, on the Dorset Coast near Swanage. This site was selected for two reasons. First, being much closer to London than Dundee, it would enhance communication between the institution, the military and civil authorities and the manufacturers. Second, the location was thought to be safe from air attack. In early 1940 this was, of course, true; the English Channel came between two allies, France and Britain. However, after the fall of France in June 1940, Luftwaffe forward air-bases were suddenly very close, and AMRE was threatened with the real possibility of air attack. Fortunately, the Germans believed that the British wouldn't be so stupid as to place a top secret establishment in such a vulnerable position, so the site remained unscathed, despite being marked by a very obvious 350 foot high Chain Home early-warning radar (CH) mast.[3] Indeed, the Germans only attacked the whole radar chain system once during the Battle of Britain.[4] They were very successful and put several stations temporarily out of operation, effectively 'blinding' the British air defence. This forced the RAF to mount standing patrols of fighters for the day that the system was out of operation.[5] However, the stations were quickly repaired and were never attacked again.

At the same time as the re-establishment of AMRE on one site, the decision was taken in the Air Ministry to expand the small amounts of work being done on centimetre wavelengths into a full research team, which would have its own hut on the new site. Work on readying the site began in late March 1940, and by April several AMRE researchers, including Lovell, were informed that this was to be their final destination. The group that was assembled there did not all arrive together, but most people had come by the second week in May. Professor H. Skinner (who had been Lovell's PhD supervisor at Bristol University) and Dr J. Atkinson arrived early, as Lovell noted in his diary on 10 May: 'Spent the afternoon with Skinner and Atkinson planning a 10 cm [split-anode] magnetron drive and a scrounging visit to Bristol.'[6] Both had been recruited to work on modifying the radar chain in September 1939. Their work done, they moved to Dundee in March 1940 and commenced their own rudimentary 10 cm antenna research.

Hodgkin arrived with Lovell in early May. Philip Dee, a physicist from Cambridge University, arrived at Worth Matravers on 11 May, together with several colleagues from his Cavendish research group, including W. E. Burcham, Sam Curran and Devons, each of whom had lectured Atkinson at the Cavendish two years previously.[7] They came from an establishment in Exeter, where they had worked on an anti-aircraft system that involved firing rockets on wires.

On 14 May Reg Batt arrived. He was one of about a dozen people who were recruited to be technicians in the new establishment. Previously he had worked for the Post Office Engineering Department, and answered a low-key advertisement in *The Times* requesting people for an undisclosed government task. On arrival he was finally told of the role of AMRE. This was explained as designing devices to locate enemy aircraft and ships using radio waves, something that seemed in the realm of science fiction to the young man.[8] Batt was described by Atkinson as being 'there to show us how properly to wire up anything electrical'.[9] Batt's descriptions provide a useful counterpoint to those of the scientists, being written from a completely different perspective – that of a technician. They often have the effect of rendering *strange* to the reader what to the scientists seemed at most problematic. Science is rarely made by the scientist in isolation. The technicians' help is often crucial, yet their presence is rarely acknowledged directly.[10]

Despite his different grade and background, Batt was accepted smoothly into the team, as his account of his introduction to Skinner and Atkinson, written in 1991, shows: 'They greeted me with genuine warmth which somewhat took me aback, used as I was to a hierarchical institution such as the Post Office, where a junior grade would be expected to know his place. It seemed I had now entered the more democratic ambience of university researchers.'[11] In this quote Batt alludes to the special relationship that operated between technicians and their scientist colleagues. The developing relationships between the various staff at AMRE, and more particularly among the centimetre researchers, are covered in more depth later in this paper.

THE KLYSTRON EXPERIMENTS

The centimetre team were allocated facilities on C-Site, which was some distance away from the main A-Site. They had a simple wooden hut, very basic with 'only one stool and a filing cabinet. There were no facilities', apart from an Elsan chemical toilet which was cannibalized by Atkinson for use as a waveguide for klystron experiments. Despite the fact that Randall and Boot's promising new cavity magnetron was already being engineered at GEC, it was still top secret, and only the head of AMRE, Rowe, knew of its existence at that time. The klystron valve, designed in America, was the only known source of centimetre waves capable of approaching even moderate power levels of around 100 watts, and this was to be the starting point as a power source for the research team. Hodgkin described their hut: 'It now seems extraordinary that all this high-powered work went on in a small isolated hut, reached by a muddy track, with no gas, and only makeshift water and electricity supplies.'[12]

A. G. Ward and Derek Robinson, both university recruits, were allocated the task of designing circuits on arrival.[13] Not everyone was treated so efficiently, with Hodgkin, for example, initially unsure of what he was supposed to be doing, 'except that [he] should design a scanning system for a 10 centimetre Air Interception set'. As his skills were not immediately required, he took to spending a lot of time at Christchurch (the nearby airfield where AMRE's test flying was conducted), where he 'learnt a good deal about testing radar equipment in a military aircraft'.[14] There were great difficulties with this, not least the variability of the power-supply frequency and the destructive effect of unpressurized, bumpy aircraft on delicate glass components. These were hazards that Lovell had already noted in the previous year. Hodgkin returned to permanent residence in Hut 40, C-Site, at the end of June.

When he arrived in May, Batt was informed of the group's activities by the deputy superintendent of AMRE, W. B. Lewis. The 10 cm work sounded the most interesting, and Batt volunteered for it. He was allocated as an assistant to Ward, who was working on the receiver. Ward explained to him the nature of the work that was being undertaken by the centimetre team. It was completely novel to the young engineer, and this was Batt's reaction to it, as he noted in 1991: 'What kind of madhouse had I elected to join? Here was I, having recently acquired formal radio-engineering qualifications, encountering alien concepts such as transit time and resonant cavities – a far cry from broadcast communications.'[15] Ten centimetre work was of a completely different nature to that which was usually done in radio engineering. The fact that the work was so out of the ordinary for radio-engineers was a contributory factor in the decision to employ scientists, and mainly physicists, on the task of designing centimetre radar. In a 1990 paper Batt put it thus:

> Facetiously it occurred to me that the powers that be had found themselves lumbered with a bunch of academics who knew nothing about radio, and so had decided to let them loose on this project which respectable radio engineers considered impossible.
>
> I was the odd man out. I had come from the Post Office, was the most junior of the group, but the only one with formal radio engineering qualifications. Yet within hours of my arrival at TRE (then AMRE) I could see that with centimetres, the less one knew about radio the better.[16]

He then explained the reason why it all seemed so strange: 'To some one like myself accustomed to working at Broadcast frequencies, it was like being told that our motor cars were useless because they could not keep up with a bullet in full flight.'[17] To underline this point, in a 1990 booklet Burcham described the reaction of the other AMRE members not engaged in centimetre research: 'The group was regarded with amused tolerance by the rest of the Establishment and was allocated a remote hut on a near cliff-top field in which to conduct their university-type experiments for which only the very far-sighted could see any predictable useful outcome.'[18] Burcham's comments highlight the reasons why the centimetre researchers were so different from the other AMRE workers. They were not civil servants, and they were not radio engineers. Nor were they working on refining existing apparatus and techniques. They were university scientists entering a new field of research. I believe that the characteristics that set them apart also facilitated the rapidity of their work, as I will show in this section and the next.

Another perspective on the events comes from Philip Dee, who kept a diary during this period. The entries written about the time of his arrival at Worth Matravers give us his viewpoint about his colleagues, the experiments they did and his battles with officialdom. In particular, he wrote comments about his colleagues which, although highly subjective, offer further insight into why centimetre research had been so badly ignored prior to May 1940. He described Rowe as 'very self-important and conceited'. Bowen, on the other hand, was 'in great disgrace over [the] scandal involving the minimum range [of 1.5 m AI]'.[19] The disorganization prevalent at St Athan earlier in the year was showing signs of re-emerging at Worth Matravers. He noted that this situation had to be rectified quickly. Dee's concern probably originated from his speaking to Lovell, whom he described as 'effervescent' and 'full of rage' at AMRE's chaos. This chaos and his opinions of the personalities of Rowe and Lewis were recorded in his diary: 'AMRE is apparently run along the lines in which separate groups working in different places work upon transmitter, receiver, display etc. This might be reasonable for established techniques but is obviously quite impracticable for a new field such as centimetres is likely to prove.'[20]

Clearly Dee had firm ideas about how the group should be run. He valued close interaction along the lines of that which occurred in his own experience at university laboratories, and in particular the Cavendish Laboratory. Unfortunately, Dee had been a superior of Lewis at the Cavendish, which meant that their previous relationship was now reversed at AMRE. In the summer of 1939 Dee, Lewis, Devons, Curran and Burcham worked together on nuclear reactions using high-tension equipment. Even though they had acquired some experience with electronics, most of this type of work was done by Lewis. According to Burcham, this was the main reason why Lewis was appointed to AMRE. He remembered that Lewis disappeared during 1939, and he was told that Lewis would be required for government work during the 1939 summer long vacation.[21] Writing in 1990, Atkinson described Burcham as 'quiet and restrained, possibly as a result of previous years spent working with Dee at the Cavendish, but he could be relied upon to get Dee going again should there be a lull – or to warn us what to expect next.'[22]

Skinner and Dee set about trying to persuade Rowe and Lewis, who were 'uninterested in AI' against their 'Pre-war civil-service mentality' to 'centre all the work in one place'.[23] Dee had also to discover how he was going to fit into the hierarchy at AMRE, such as there was at that time. The reversal in the power relationship between Dee and Lewis had the effect of subverting the formal lines of authority at AMRE as they applied to the centimetre group. Consequently, the members of the group often acted outside the regulations of the establishment, but in the belief that they were circumventing restrictions to rapid progress. This behaviour was tolerated grudgingly if it produced results.

The relationship between Dee, Skinner and the leaders of AMRE was recalled by Burcham in his 1990 booklet. He remembered that Dee and Skinner were officially group leaders under Lewis. Bowen, antagonistic to Rowe and his cohorts, saw Lewis as an active hindrance towards progress. On the other hand, Burcham believed that Lewis pushed for the centimetre work to be continued during the Battle of Britain and threat of invasion, when all 'non-essential' work was under heavy pressure to cease. If this was the case, then he had a positive effect towards ensuring the continuation of the project, at a time when many people, not directly connected with it, thought that the researchers would be better employed on more pressing problems.

Skinner, who came from Bristol University, had a similar level of experience and status to Dee. At the beginning it was unclear whether he or Dee was in charge of the centimetre work, although each had his own area of responsibility. Dee was the one who made arrangements and offered leadership. He was a tidy, meticulous person, who was well organized and 'loved an argument'.[24] People went to him for guidance. Skinner was more of a 'virtuoso', who dipped into various projects, including using the split-anode glass envelope magnetron to provide a power source for the aerials. He used his connections to acquire low-power klystrons from the Admiralty valve-builder Sutton at Bristol University to be used as local oscillators in the receiver. He was remembered as being 'rather quiet and self-contained ... in any discussion you had to decide whether the grunts he emitted were in favour or not of the views being expressed – and in age he was the senior member of the party.'[25] He built crystal detectors with a considerable degree of skill, carefully fitting the crystal on to the cat's whisker wire, and tapping it carefully with his pipe.[26] It was a delicate job, and Skinner took until July to perfect the process: 'Skinner made v. nice glass sealed crystal, the great advance being his discovery that Si [silica] can be brazed to rod. This enables glass seals to be made without disturbing crystal mounting.'[27] The process was also described by Batt, who, as a technician, appreciated Skinner's practical abilities: building crystal detectors was 'a task taken on by Skinner ... who had not only the vision of what was required but the practical abilities, including glass-blowing skill, to make [them].'[28] These two quotes illustrate two points. First, Skinner was highly skilled in a variety of practical arts that enabled him to construct the apparatus and components necessary for his experiments. Second, both quotes show that Skinner's practical ability was a part of the discovery process for him. His ideas about how to proceed were shaped by his understanding of how to *make* equipment. Although Skinner was probably the most skilled practical experimenter in the group, his abilities were by no means unique. All the other members of the team had similar experiences and skills, if to a lesser degree, which actually shaped the direction in which they went. They constructed their apparatus and developed their knowledge of how to use it through physical interaction with it.

At the time of his arrival, Dee was informed by Rowe of some of the work going on at Birmingham University:

> Oliphant is said to have developed a klystron giving a kW at 8 cm and Randall also in Birmingham has a magnetron on the same wavelength. There are no transmitting valves at all in AMRE on such wavelengths and Skinner and I have decided that I should go to Cambridge, and try to get a klystron made in the Cavendish to Oliphant's design as quickly as possible.[29]

Other sources (such as Lovell, *op. cit.*, note 2) suggest that the magnetron was a complete surprise to the centimetre team when it arrived in July, but perhaps Dee, being more senior, was informed earlier and Lovell was not aware of this.

The quote in the previous paragraph shows how Dee's Cavendish Laboratory connections were of great use to him. Another ex-Cavendish member of the group was Burcham. When interviewed, he shed some light on the working practices used at the Cavendish that were also adopted at AMRE owing to the high proportion of former laboratory members there. At the Cavendish people worked independently, on separate projects, but regularly interacted with each other to discuss their day-to-day problems. This way of working continued at Worth Matravers, where the layout of their hut mimicked laboratory conditions, facilitating communication between the researchers. It was a long hut with a bench down each side. There was an office at the end shared by Dee and Skinner, who were not very compatible, as Skinner's untidiness infuriated Dee. Further evidence of the way people in close confines can interact was given by Dee's comment in his diary entry for 5 July that Skinner's general untidiness was a problem for Robinson, to whose group he and Dee belonged. Burcham, however, made note of Dee and Skinner being incompatible, so perhaps this was how Dee vented his irritation with Skinner – by blaming it on someone else. The degree of freedom of communication allowed within the team is surprising in the light of their secret military work, but Burcham believed that it contributed greatly to their rapid success. The whole group was extremely close-knit. Regular interaction was of great importance. They kept little paperwork, apart from technical references, and generally wrote very little down. This was confirmed by Dee when he made note of writing a report to Rowe, as I show in a later quote.[30]

Over the next few weeks after their arrival in mid-May, Dee set about initiating work on 10 cm. On 21 May he went to Cambridge to see Bragg (the head of the Cavendish Laboratory) about getting a klystron built there. On 22 May he went to Birmingham to see their klystron and also, according to him, the cavity magnetron. On 23 May he went back to Cambridge to check on their progress with the klystron. On the same day, Lovell 'started active attempts to get a 10 cm [split-anode] magnetron going in the field.'[31] Meanwhile Dee was confirming how he fitted into the structure, and recorded that on 27 May he had a meeting with Rowe to determine where and under whom he should work.

By the end of May conditions appear to have improved in Dee's view, as on 30 May he 'moved into Hodgkin's aerial hut and Burcham and [he] put together some HT gear for the klystron'. The next day, 'Burcham and [he] had a fine day putting up pumps, getting 3-phase off the mains and running the high vacuum backwards. Quite like old times.'[32] As Dee and Burcham had already worked together in the Cavendish, it is likely that they would have established a rapport that allowed close co-operation. This is an example of how they sought to recreate the working environment that they felt suited them, and that they believed increased the speed with which they tackled their problems.

While Dee and Burcham worked towards setting up their klystron,

Lovell noted on 29 July that his split-anode magnetron was showing signs of working, and that Dee was building a klystron inside the hut. Lovell's magnetron was 'going on 11 cms' on 30 July, which was a big step towards him being able to commence his aerial experiments.[33]

Early in June Dee made mention of the way he was now fitting into the establishment, of his views of Lewis and of what he was now supposed to be doing: 'Lewis ... seems to be a little hedgy perhaps for fear that Skinner and I are showing signs of absorbing too much work and perhaps becoming too powerful.'[34] This quote shows how, even with a pressing external threat, and working in wartime, social relations within the group could have an effect on that group's performance. This is reinforced by Dee's note that he also pressed Lewis to tell him what exactly he was supposed to be responsible for. Dee and Skinner, being the senior men in the centimetre group and with no official demarcation between their activities and spheres of influence, were starting to tread on each other's toes. Skinner was deemed by Lewis to be in charge of basic research, and Dee was made responsible for putting the system into an aeroplane. However, Dee couldn't see this happening for at least 6 months, and Pringle (of the 1.5 m team) had been given that particular job. As he lamented: 'as far as I can see all that I am really supposed to do is ask Pringle to get it [installation of centimetre AI] done. This should not take six months.'[35]

Dee's diary gives an impression of considerable chaos, which was the case during the first few weeks. However, despite this chaos, some of the team were managing to do some research on their project. While Dee and Burcham were working on completing the klystron set-up, 'Lovell ... got a low power 10 cm glass envelope magnetron working in a "field" coffin for the polar diagram measurements ... much to Lovell's annoyance everyone twiddles the knobs on every possible occasion, this being the first working apparatus.'[36] Dee's description gives the reader an impression of a cosy, familiar laboratory where everyone is interested in everyone else's work. Such an atmosphere would have been very similar to the laboratory from which he had only recently come, and highlights the difference between the actuality and what one might have expected from a group of government employees engaged in secret work. This impression of chaotic endeavour is reinforced in a letter by Dee from 1982: 'I think that you will never appreciate how unorganized [sic] and chaotic the whole set up was. We went our separate ways but seemed to coalesce when there was a common agreement – or enemy!'[37] Often at this stage the 'enemy' was GEC, not the Germans. GEC was working on a separate Air Ministry contract for a 25 cm system using conventional glass-envelope valves. GEC's comparatively greater success at this stage, which was being shown to influential people in the Air Ministry and the Government, meant that AMRE's own work was under threat.

Lovell's notes confirmed that he was able to make polar diagram measurements of the horn antennae fed by the glass-envelope magnetron acquired from Bristol on 5 June.[38] The previous day he had made no measurements because 'Atkinson was messing about with pulsing and

Skinner with crystals.'[39] Skinner and Ward were working on the receiver, using a klystron from the Clarendon Laboratory at Oxford University as a local oscillator (as opposed to the klystron being built for Dee, which was to be used as a source of power). This became known as the Sutton tube after its inventor, Dr R. W. Sutton.[40]

The urgent nature of their work, and the twin threats of German invasion and GEC's rival system, led to a rapid pace that favoured continued experimentation. The need to work quickly meant that the majority of learning that was done by the researchers was passed on directly through close co-operation within the laboratory hut. There was not time to write up results other than as brief notes in laboratory notebooks. However, the other aspects of university life, such as the seminar, did make an occasional appearance. Dee noted that on 10 June he went to a lecture given by Bowen, the second such lecture in fact, on 1.5 m AI and features he believed would be required by centimetre AI. He described Bowen, who clearly impressed him, as a 'terribly competent man'. It is also possible to infer from this quote and his hostility towards Rowe and Lewis that he was also unimpressed that Bowen, a man with four years' airborne radar experience to offer, was being excluded from playing a full role in their work.

In the entry for 10 June he also mentioned their continued attempts to conjure up something from the transmitter klystron. A large part of the delay in their experiments was that they had to wait for a water supply for cooling to be rigged up. Either owing to the level of confusion in the new establishment, or because of the slowness of official channels in a government establishment, nothing was done for several days after their request was made. In the end Dee went to Cambridge and 'stole' the necessary equipment. Despite his attempt to speed up the process by using this unorthodox method of acquiring equipment, it still took seven men ten days to rig up their booty. In disgust, Dee wrote that 'the whole thing could have been done by Lincoln at the Cavendish in an hour'.[41] While he does not state so explicitly, it would appear that this is a reference to his laboratory assistant from Cambridge. Such references from scientists are rare (although this was written in his diary, not a source that was intended for publication) and it hints at the intimate relationship, confidence and understanding that the scientists had with their technicians. It also indicates that part of the reason for the friction at AMRE between the university scientists, such as Dee, and the civil servants, such as Rowe, was their completely different methods of working. These frictions were exacerbated by the unusual circumstances of high pressure to achieve results quickly, caused by the war situation. This pressure caused the peacetime structures and methods used at AMRE to be put under severe strain, and their consequent circumvention by the centimetre group.

Ward and Batt moved into the hut during mid-June, joining up with Skinner, Dee, Burcham, Lovell and Chapman, who had moved in slightly earlier. Batt was initially amazed to see the klystron equipment, which was 'reminiscent of those early Frankenstein films'.[42] Their klystron had come from drawings furnished by Mark Oliphant, head of the group at Birmingham University. He had visited the Hansens and seen their original klystron, constructed at Stanford in the USA, in early 1939. Unfortunately, their klystron was very temperamental, and the AMRE version was proving no less so.

On 12 June Lovell started working on parabolas instead of the horn antennas, and noted the results in his laboratory book: 'moved the dipole across the mouth of the sectional parabola and found that it shifted the beam 8° per 5 cm ... not enough power to see what the side lobes were like ... this makes me regard the aerial problems as 75% solved.'[43] When Lovell's valve failed the next day, he phoned up Megaw at GEC (who was also modifying Randall and Boot's cavity magnetron into an engineered form, although Lovell did not know this) to see whether another could be found. Thus, despite the developing antagonism between GEC and AMRE, there was still evidence of some beneficial co-operation between the two organizations.

This (13 June) was also the day when the team finally produced power from its klystron set-up. Up until then the klystron had been incredibly difficult to operate, just as the Stanford klystron had been. As Batt put it, 'It seemed that although the theory was simple enough, for the klystron to build up its powerful oscillation every facet of its construction, assembly and adjustment had to be just right. Here was the bane of many a scientist – an entity possessing an infinite number of variables.'[44] Batt's quote provides good evidence to back up Gooding's assertions[45] that theoretical knowledge is not sufficient to conduct experiments. The experimenter must also possess practical experimental skill: 'every facet of its construction, assembly and adjustment had to be just right'. In other words, the experimenters had to have spent time working with their equipment, absorbing practical, unarticulated 'know-how', in order to get it to work.

Burcham also noted that the klystron first oscillated on 13 June.[46] One of the main problems with it was that the filament often burnt out and had to be replaced, which meant dismantling the valve as Dee described in his notebook:

> Whenever I am out of the lab and Skinner has to do this he forgets to turn off the water before pulling off the cooling pipes with the result that I am standing all day in about 1/2" depth of water, and the water on the bench is about equally deep but has its surface relieved somewhat by floating cig-ends, tea leaves, banana skins, etc. However, we have managed to get quite a lot of power out of it and light a Pea lamp nearly a foot away.[47]

The klystron remained difficult to use until the cavity magnetron arrived in July, which was the main reason why the cavity magnetron was so quickly employed. Burcham noted, as late as 15 July, that he had 'Tried pulsing of klystron again. Got it going, but it then suddenly stopped & would not work for sometime. Finally got it again by heating cathode v. hot.'[48] This quote shows the learning process that Burcham was going through with the klystron. He was still not in a position where he fully understood it, but he had developed ideas about how to fix it when it went wrong or refused to operate. In other words, he had already learned enough about klystron operation to know what type of remedies might succeed in getting it going again. But the klystron was clearly far from unproblematic.

Throughout June work continued in the hut, where a routine was developing. Dee, as we have seen, gave a very different impression of life in the laboratory in his diary notes than would normally be found in experimental descriptions published in scientific textbooks or papers. Nevertheless, we are able to follow the order of events, and are offered personal colour by examples such as this: 'Lovell is now measuring polar diagrams on 10 cm in the field outside the hut. His mirrors [parabolic aerials] provide excellent targets for competition between Atkinson and myself of the throwing of large lumps of mud which collect outside the hut.'[49] One can only imagine that this must have irked Lovell considerably!

Lovell was being assisted by Chapman, the technician who had accompanied him from Manchester. Batt pointed out how well Lovell and Chapman complemented each other:

> In Chapman, Lovell had a most valuable asset. A university laboratory assistant was truly a jack of all trades, and master of them all. He combined the abilities of an instrument-maker, a tool-maker, a glass-blower, and an expert in vacuum technique. In addition he would have a working knowledge of electronics, optics and thermodynamics. To a university laboratory a good assistant would be as valuable as a master chef to a maître d'hôtel.
>
> Chapman was all of these things but above all he was by temperament calm and unflappable and in no way intimidated by Lovell's hyperactive zeal. Hence they made an effective and extremely productive team.[50]

Batt, as a technician, would be expected to be more aware of the assistance that such people give in the laboratory. In practically every other account given by a scientist, technicians are almost invisible: rarely, if ever, alluded to. Here we can see precisely how much of a contribution they could make. This is not to say that scientists did not also have these skills, for some of them did, notably Skinner. But there was definitely a partnership and division of labour occurring.

On his return to Worth Matravers, Hodgkin was allocated the task of designing a scanner for the system:

> Someone, probably Dee, suggested that I get a prototype scanner going, first on the ground and then in an air-to-air test to take place by Christmas [1940]. This was a tall order, as we then had no way of transmitting and receiving on the same aerial. Nor did we know the answer to much simpler questions such as how to make a satisfactory rotating joint in a concentric cable or wave-guide carrying electromagnetic waves at centimetre wavelength.[51]

In this paragraph Hodgkin uncovered the problems that were facing every member of the team in one way or another: they were working to a very tight schedule, and very often did not know whether they would be able to meet it. However, this pressure forced them to take risks which, surprisingly, often paid off.

Dee's privileged knowledge about the new cavity magnetron was increased when he learned on 20 June, from a letter from Paterson (Head of GEC) to Rowe about GEC and AMRE co-operation, that the new magnetron now gave 2 kW of power. This was ten times greater than anything else on this wavelength, and was a significant breakthrough. The others were unaware of this, but Dee now viewed their klystron work solely as a chance to learn about centimetric wavelengths and their behaviour, rather than as the main thrust of research:

> With the klystrons we are learning the elements of aerials and transmission lines, much to the amusement of a fat old man called Bartlett ... he knows all about the subject but regards it as quite pointless to let us into the secrets. Tried every ... way to get some answers from him but gathered that I shall not be ready to understand anything for another 2 or 3 years since the subject is so complex.[52]

Despite Dee's sarcasm, one can see the gulf that existed between the majority of the AMRE staff and the strange group at C-Site who persisted in their odd experiments that others thought were likely to lead nowhere. Support for this view of the existence of a gulf comes from Hodgkin, who commented in 1992 that a 'curious feature of the build-up was that Dee's team was largely composed of physicists and we had hardly any mechanical or electrical engineers'.[53] I noted previously that this meant that the researchers did not have any preconceptions about what was possible. Certainly, when we see what Batt wrote earlier about centimetric research seeming very strange to a radio engineer, this appears to be the case.

On 26 June Rowe finally confirmed Dee's status as being in charge of installing 'AIS' (centimetric, S-band (10 cm), AI) into an aeroplane. Dee complained again in his diary that this was highly unlikely to occur for a further 6 months. Further discussions with Rowe on 12 July resulted in Dee being told he was actually in charge of all centimetre research. This elevated Dee to a position superior to Skinner. It went some way to removing the friction in their relationship, as the two had hitherto been treated with equal seniority by the more junior members of the team. Later the two went into different spheres, with Dee put in charge of applications, and Skinner of basic research on 9 cm and 3 cm.[54]

The AMRE group's relative lack of success until the end of June was in contrast to the success of the researchers at Birmingham University, who had worked with the klystron since the previous October. They had also, of course, developed the cavity magnetron. Their klystron equipment and transmitter was brought down on 3 July by Oliphant and Sayers and their group, and demonstrated to the AMRE staff, as Dee noted.

Burcham, however, thought it was on 2 July. He informed me in interview that it was far more refined than the AMRE klystron, but still worked using a pump. As far as I am aware, this was the first ever working 10 cm system – but there is no mention of it anywhere apart from in Burcham and Shearman's 1990 description. I suspect that this is because of the subsequent success of the cavity magnetron powered set produced by AMRE, which would have tended to sideline the unutilized klystron equipment from Birmingham in the minds of the AMRE researchers.[55]

When compared to the Birmingham team's success, AMRE's klystron equipment was still going nowhere. Atkinson wrote:

> The project [was] in difficulties. With our relatively massive equipment, any number of variac transformers, Avometers and oscilloscopes, the idea that even if we were able to get echoes at a reasonable range, we would be able to fit it into a fighter aircraft seemed most unlikely, not only to visiting Service personnel, but also amongst many of our own colleagues working on other projects.[56]

This quote indicates that, as time progressed, confidence in the AMRE team's ability to produce a centimetre AI was decreasing. In order for them to survive, they had to produce *demonstrable* results that were as good as, or in excess of, those being produced at GEC and at Birmingham.

Meanwhile, Dee was unhappy with the co-ordination of effort on the development of centimetric equipment. On 9 July he had a meeting with EMI, who had been contracted by the Ministry of Aircraft Production (MAP) to work on valves, aerials and receivers: 'Presumably Warmsley at MAP is the only co-ordinator between the firms and AMRE and he never touches a piece of apparatus, just tries to direct the work from an office desk in London.'[57] Despite being aimed at bureaucratic interference and incompetence, this is a revealing comment, as it further underlines the implicit belief by Dee that the scientists had to get to grips, literally, with their equipment in order to realize their full potential. Dee's concern about the situation at Standard Telephones and Cables (ST&C), another subcontractor, which was worse than at EMI, also illustrates his implicit belief in the need for close co-operation between all the parties concerned with development. He described ST&C as 'obviously need[ing] intimate contact with the other centres of such work'.[58] Again, on 13 July 1940, he was very disparaging about a bureaucrat: 'D[irector of] C[ommunications] D[evelopment – Sir George Lee] is here for a meeting on AIS [10 cm AI]. He is a hopeless old man with no knowledge of any of us.' Dee firmly believed that there were very great problems with communications between the higher echelons of the Air Ministry and the work that his team were doing on the ground, as a 1981 letter shows:

> The brutal fact was that in early 1940 people at MAP were thinking they could develop a 10 cm AI by sending secret, limited information out in different directions – to firms, to RAE, to TRE etc., etc. The apparent idea was that all this information would come together at

> MAP and an equipment evolve. It was just ludicrous and neither Skinner nor I would take part in such a game. Instead we went ahead to make an equipment, pulling in firms as required.[59]

Dee and Skinner's resistance to this policy of what they saw as excessive secrecy shows that they believed that close co-operation was essential for rapid progress. They realized that it was important for the people actually building and experimenting with the equipment, and those who would manufacture it, to be closely involved at all stages of the process. However, at first they had to pursue this policy against the inclination of the organizations with which they worked. It was a measure of their status that they were allowed to work in this fashion. The policy of openness was eventually enshrined in Rowe's weekly 'Sunday Soviet' meetings, which brought together researchers, airmen and politicians for free and frank discussions about AMRE's work. It is also interesting to note that the process that they describe as being so detrimental, the 'sending [out of] secret, limited information ... in different directions', was exactly the process used in German radar research, and which had a detrimental effect on the speed of their developments.

One of the effects of the centimetre team's policy of autonomy was that the researchers were expected to take responsibility for the equipment they were building, as Hodgkin related: 'In 1940 when I was still a junior scientific officer earning £300 per annum I was encouraged to write and sign letters to firms on technical matters, and Dee's authority was required only if some point of major policy were involved.'[60] As the system was also subject to a weekly review, this stopped over-enthusiastic young researchers from going too far, as Lovell did in the winter of 1940. He ordered several thousand square feet of Perspex in order to construct a covered 'greenhouse' within which to perform aerial experiments without being exposed to the elements. He reasoned that Air Ministry procedures were such that ordering a large amount of material was as easy as a small amount. When this misdemeanour was discovered, Lovell was reprimanded for overstepping the mark, but it was further evidence of the clash of cultures at AMRE.

The result of Sir George Lee's visit on 13 July was that the relationships within the 10 cm team were unofficially formalized. Burcham recorded this relationship in his notebook:

> I worked with Atkinson on pulsing the klystron, with moderate but not especially great success. A family tree has now been drawn up for 10 cm work on the following lines (unofficial):

Dee

(T)	(R)	(Aerials & Presentation)
Skinner etc.	Ward etc.	Lovell etc.

> DCD is coming tomorrow to give some decision about the priority of AI.[61]

Burcham's mention of further klystron problems shows that the visit of Oliphant and the Birmingham klystron system had not, at this point, led to any improvement. This lends weight to the possibility that the klystron was not fully understood by the Birmingham team, if they were unable to assist the AMRE team in making their klystron work.

Now that he was (unofficially, at least) in charge of the 10 cm project, Dee's frustration at the perceived bureaucratic forces ranged against him began to mount. Being in charge of the project exposed him to higher levels of authority than he had experienced previously, as described in the following passage from his diary:

> There has been a lot of talk about the impracticability of centimetres due to specular reflection. All the defeatists quote this as a reason for not driving for centimetres. Even at high level meetings in London between people who have never seen a centimetre valve these objections are being aired. Some people believe that this is likely to make the pick up very uncertain.

This is an extremely significant quote. The main reason why the Germans failed to pursue centimetre waves was on the basis of experiments conducted in 1939 which led them to believe that the reflections from aircraft illuminated with centimetre radiation would be too weak to detect. Their experiments led them to construct a theoretical explanation that confirmed this view, and, once established, this view informed all further experimental work until the arrival of the British H_2S.[62]

The distance of the officials from the experiments at AMRE led to a conflict between their theoretical beliefs about centimetres and Dee's beliefs, informed by his experience of actually working with these waves. The AMRE team's practical knowledge of what they could actually do led them to challenge the view that centimetre waves were useless. It would have been quite possible that the 'official' view would have been the one that pertained, as happened in Germany where authority carried a lot of weight. However, rather than just accept these 'facts', the AMRE group decided to investigate for themselves, as the remainder of the quote from that day in Dee's diary illustrates:

> We are therefore doing a rough polar-diagram of a Lysander with 50 watts of 8 cm C. W. and 6' mirrors, as was pretty obvious the aircraft had a pretty peaky polar diagram but at all aspects there seems to be a fair return. Hope this will allay the boneheads at the top. Had to write this result formally on a file for the Chief Superintendent [Rowe] to read![63]

Dee recorded the results of these experiments in his diary in a very interesting fashion, as he did not use standard scientific convention. Using the split-anode magnetron that Lovell had for testing parabolas as a power source, they mapped the reflection pattern of centimetre waves from a Lysander aircraft. The 'peaky' pattern which they noted referred to there being more energy reflected at certain angles than others; for example, the side offered a bigger reflective surface than the front. However, 'there seems to be a fair return' is not the sort of language one would expect in a report or paper. This indicates the still uncertain nature of their beliefs about centimetre radiation at this stage. Furthermore, that he should make special mention of having to write up results formally indicates that it would not have been a usual practice. This was also confirmed by Burcham, who, in interview, said that it was 'an effort to write anything'. He also said that there were no progress reports in general: brief notes were written down for personal use, as *aides-mémoires* and to develop ideas in a graphical form. Equipment was improved through trial and error, as they conducted experiments.[64] Their mode of working was one of close, informal co-operation in a highly practical environment.

By mid-July, 2 months after the arrival of the team members, a formal structure to the group finally emerged. At the same time the cavity magnetron arrived at Worth Matravers. The arrival of this valve altered the speed of progress significantly. Hitherto they had managed only to perform some useful experiments to increase their understanding of the behaviour of waveguides, parabolas and other centimetre components. Their failure to stabilize the performance of the klystron so that it gave them a regular supply of high-power, frequency-stable centimetre waves had significantly retarded the development of a complete 10 cm system. They could not begin to combine the components into a complete, *demonstrable* system if they were not able to operate one of the key components, the transmitter, consistently and to order.

THE MAGNETRON ARRIVES: THE INITIAL SUCCESS ON 10 CM

On 20 July 1940 Dee noted that they received the GEC magnetron (referring to the E1189 sample redesigned by Megaw from Randall and Boot's original), and that he and Atkinson fixed up a modulator to drive it. Burcham noted that the 'magnetron [was] imminent' on 18 July, and on 21 July he mentioned using it. However, he recorded it as coming on Friday 19 July, and that it was brought down by Atkinson.[65] Batt wrote that on 19 July Atkinson cleared bench space for 'a new and exciting toy'.[66]

The magnetron had an immediate effect on the 10 cm group. All the work done at GEC by Megaw had made it into a device that was far simpler to operate than the klystron, which was still prone to stop without warning. Megaw's skill was encapsulated in his version, which no longer required the same level of skill to operate as when it had been in Randall and Boot's laboratory. By contrast, the klystron was not fully understood, and the AMRE scientists did not have the skill to operate it. With the new device, the AMRE researchers had to learn neither to produce the effect,[67] as did Randall and Boot, nor to replicate[68] the device (as this had been done by Megaw). They only had to 'switch it on'.

The small, neat, compact magnetron, 'the size and shape of a pill box', was fixed up in a wooden cradle between the poles of a permanent horseshoe magnet. Cooling air was supplied by a large blower (as opposed to the klystron's network of cooling water pipes). The high-voltage supply for the klystron was now used for the magnetron. Batt described what happened when the setting up was completed:

> With the precious magnetron in position and connected up, Jimmy [Atkinson] began to wind up the high-voltage supply. One thousand, two thousand . . . nine thousand, ten . . . Suddenly a neon lamp being held against the output stub of the magnetron started to glow: the magnetron was oscillating, producing power – 10 centimetre power! It was all so simple, unbelievably simple. We didn't audibly cheer but nevertheless that was the feeling we all shared.[69]

They now had a power supply that was stable. It was one less thing that they had to learn *how* to operate.

On 2 August Burcham noted in his diary that 'The magnetron is now ready to be put outside for reflection experiments. I am supposed to be in charge of this although whether Atkinson will let me be, I don't really know.'[70] In the two weeks prior to this, Lovell had moved on to measuring the transparency of perspex to radiowaves, as Dee recorded on 22 July. Lovell himself recorded that he had got 'Very good results with the 3' paraboloid. Side lobes too small to measure with the available power and marvellous beam shifting $\pm 25°$ with a shift of 10 cms and the diagram not gone too badly to pieces.'[71]

After these satisfactory results Lovell was now starting to think of using a beam-shifting method rather than a movable antenna. Such practical thoughts were happening elsewhere too, as on 29 July Dee recorded discussions with Gossling and Paterson from GEC concerning the division of effort to produce a working 10 cm AI. The issue of which AI would succeed (GEC's or AMRE's) was now coming to a head, but would not be resolved for another month.

In Dee's diary the month of August opens with 'First cm echoes in trailer at Worth'. This was the month when the team made rapid progress after receiving the magnetron. However, Dee still felt impeded by the bureaucracy surrounding him. On 8 August Sir Frank Smith (the new head of CVD) visited and was shown 'one of the few things well organised'. As I have already mentioned, the fact that GEC was making rapid progress forced AMRE to consider how it too could produce a demonstration of an apparatus. For the team to demonstrate something successfully, they had to be able to operate it as and when they required. In other words, it could not be something like the klystron that was not understood and was prone to 'misbehaving'.

Smith asked Dee what more was required for AIS: 'I [Dee] said that all that was necessary was for me to have 10 minutes with someone who had the power to tidy up the appalling muddle and lack of direction of the work.'[72] As Smith *was* that person, Dee went on to impress upon him the need for further co-ordination of the whole effort between AMRE, the services and the manufacturers. He saw AMRE as the research *centre*, the hub that connected the RAF, which used the device, and the manufacturers (such as GEC and EMI), which made the equipment that AMRE developed.

Despite Dee's difficulties with administration, things were starting to improve with the magnetron equipment in early August. The team were now at the point where all the various components were ready to be put together into one apparatus in order to start conducting tests. The transmitter and receiver were attached to different, independently mounted paraboloids, and according to Lovell, on 8 and 9 August they managed to receive clear echoes from a coastguard hut some 500 m distant.[73] However, Burcham, in his notebook, recorded this as happening on 10 August. He also made note of a difficulty: 'Running magnetron and looking for reflections. Able to get coastguard hut easily and well but difficult to direct large paraboloid onto a plane. Didn't get it in fact.'[74] Burcham's note shows that on 10 August they were still unable to correlate a signal on their display with an echo from an aeroplane. Such a calibration was very important for them. Over the previous five years, radar researchers had used demonstrations of receiving aircraft echoes to reassure their military superiors that they were spending their money on a worthwhile project. The AMRE team, facing the challenge of GEC's 25 cm system, were now also in the position of having to produce a demonstrable apparatus in order to continue their research.

The episode that follows is very important, as it illustrates several key historical points. First, it shows how Lovell, Burcham and Batt calibrated their equipment so that they were able to demonstrate its capabilities with confidence to visiting dignitaries. Second, it shows how Lovell, Batt and Burcham differ in their recollections of what happened, despite consulting their own written records of the events. Their interpretations of what happened vary according to their participatory role in the events. Third, it shows how the misunderstandings and hostility between GEC and AMRE created their own different perspectives on the personalities involved in the events, and highlights the potential for a lack of co-operation that could have had far-reaching effects.

On 12 August Dee was away from Dorset visiting GEC, which, he thought, 'fear[ed] losing control' of the centimetre AI project. Skinner and Atkinson accompanied Dee to GEC. Paterson, head of GEC at the time, had this to say of the meeting:

> What a day! Lewis, Dee, Skinner, Bartlett, Atkinson, Ward from Swanage to talk AI and means of co-operation. Except for Lewis there was no articulate idea on the subject from their end. They appear to me to be a group of individualists with little experience in team work – except Lewis. Pressed them to have fortnightly meetings with us to review branches of work and detailed programme. Dee was non-committal. I thought it best to put our proposals in writing in a letter to Rowe. I don't believe they have yet got the idea of sharing responsibility with us.[75]

From Paterson's perspective it was Dee and AMRE that were in the wrong and being disruptive. GEC had its 25 cm system and felt that it should have its achievements recognized. This was not the view of people at AMRE, as confirmed by Burcham, who wrote on 13 August 1940 that 'Dee & Skinner [were] back very depressed from GEC. There is obstructionism there.'[76] AMRE needed to be able to demonstrate their achievements if they were to head off the threat from GEC.

In a relatively quiet atmosphere, Lovell, Burcham and Batt fixed up their equipment to conduct more trials. In correspondence with Burcham, Lovell quoted from his own diary to confirm the correct sequence of events:

> Mon. August 12th: Skinner, Dee and Atkinson away so got the double paraboloid swiveller out of the workshops and with Burcham fixed up cables and by 6 p.m. we were trying to get reflections from aircraft. It was really rather exciting and we definitely picked him up.
>
> Tue. August 13th: Dee quite incoherent after meeting with GEC yesterday. News of our echoes would have been good for GEC one imagines. We had quite a successful day picking up a plane in the morning and a tin sheet and a bicycle on the hill, which was amazing considering it should be right in the ground returns. Watson-Watt and Rowe down in the afternoon and suitably impressed. We got a Blenheim over for Watts at 7 p.m. but unfortunately got nothing. Must have been because it was going tail and front on and not across.[77]

Lovell's comment that 'we definitely picked him up' is ambiguous – does it refer to Batt or the aeroplane? In his book of 1991, Lovell decided that they 'saw' the aeroplane *first*, on 12 August. However, in *his* 1991 book, Batt believed that they were *unable* to see the aeroplane on 12 August, and he had to make his bicycle ride in order for them to pick anything up: 'it was early afternoon and a spirit of depression was abroad since once again our efforts had come to nought. What was needed was a moving target that could be called up to order.'[78] Batt highlighted what the main problem was: according to him they were unable to see anything at all with their equipment and required him to cycle with a tin sheet strapped to his bike in order to provide a slow-moving, easy to 'see' target. This would provide ideal calibration for Lovell and Burcham. Batt wrote that Lovell believed he had received echoes from Batt and his bicycle 'up to saturation' when Batt performed his ride.[79]

According to Batt, it was the next day, *after* he had made his calibratory cycle ride that:

> Four very dejected physicists arrived back from their meeting at Wembley. It transpired that GEC had won a moral victory. Whilst Dee and Skinner had little in the way of positive results to report, GEC had taken them up on to the roof of the Wembley building where their 25 centimetre system was set up in a hut. Here they were shown an impressive array of echoes from the surrounding topography.[80]

Batt's quote shows how important it was for AMRE to be able to demonstrate the 10 cm as GEC had done. Demonstration provided an effective tool to use in settling the dispute over who should have control of the AI project. However, the previous day's events were to be very useful. Batt then recalled that Lovell mentioned that he had clearly 'seen' Batt on his bike the previous day. On 13 August Batt had to repeat the demonstration by getting on his bike and cycling off so that Lovell could demonstrate the echoes received from him. He subsequently repeated this for several other visitors, including men from GEC. He explained that the reason why he was being used was because he was a stable, easily 'seen' entity: at the time they only had cumbersome, independent paraboloid mounts.

Lovell described Batt as a 'junior assistant', someone who was not as senior, in his eyes, as a scientist such as himself. He also recorded in his diary that the twin paraboloid mount arrived on 12 August, which is in contradiction to Batt's recollection. This earlier date for the arrival of the twin-mount is supported by the entry in Burcham's notebook for 12 August:

> Aft[ernoon] – Lovell's double paraboloid arrived, mounted on stand. Set it up, made rough connections everywhere and looked for echoes with Ward's crystal mixer receiver. Found coastguard hut alright though there is a big ground return & echoes are very wide. Had a plane over and thought we saw it, but couldn't be sure, seemed to keep jumping in & out. Easily picked up Batt cycling around with a sheet of tin. This is distinct from ground very clearly.[81]

Burcham's quote explains the discrepancy between Batt and Lovell's accounts. Batt, as the cyclist, overemphasized his role to the point of excluding seeing the aircraft *at all*. This was not the case; they *possibly* saw an aircraft. Lovell, however, privileged the aircraft as the thing 'seen' first. Burcham noted that Batt was far easier to see than the aeroplane, which they could not be sure they had positively identified. Lovell, after fifty years, favoured the interpretation that he had definitely seen an aeroplane; Batt, that it was him that was seen. Looking at their original notes opens up the ambiguity that existed at the time.[82]

What we can also see from all three men's quotes is that the business of learning to 'see' with the apparatus was *not* straightforward. They had to move from observing something fixed (the coastguard hut), to something slow-moving and easy to direct (Batt), to something far less controllable and harder to 'see' (an aeroplane). Lovell was positive about seeing aeroplanes, but Burcham 'couldn't be sure'. Batt thought they couldn't see them at all. Despite their certainty that they could at least 'see' some things, like the hut and Batt, even these may have been artefacts of the equipment. Burcham was challenged on this matter the next day: 'Osborne says that we were looking at back kicks yesterday ... The back kicks are due to the receiver saturating. Output stage has to be modified.[83] 'Seeing' with their equipment, especially their preferred target (aeroplanes), was a difficult business. However, as I have mentioned, they had to have some results in order to keep their project going owing to the threat from GEC and from more sceptical members of their own establishment. This is where Batt came in as useful: he was an easily directed demonstration aid. The necessity of having a good demonstration is confirmed by Atkinson, who wrote that

> [Rowe] could not understand the abundant enthusiasm shown by Dee and Skinner and their complete assurance of success. You must remember that France had just fallen and I think Rowe was under some pressure to concentrate efforts on projects likely to assist against more immediate air attacks – and there were continual rumours of possible invasion.
>
> My impression that the future health of 10 cm work was not 100% was confirmed when I ran into John Mercer, whom I had known well at Bawdsey and who worked directly as personal assistant to Rowe. Amongst his duties was the arranging of high-level visits and he told me of a forthcoming visit and said that we had better make certain our echo demonstration worked well. He also bet me £1 – a lot of money in those days – that 10 cm would never be used operationally during the war.
>
> My personal opinion is that the quiet but very strong and definite influence of Ben Lewis was a major factor in the group's being given more resources, now that satisfactory echoes had been demonstrated. A second factor was the support of Air-Vice Marshall Joubert of Fighter Command. Joubert came . . . for a demonstration following one of Rowe's famous Sunday meetings: the demonstration went well much to our relief.[84]

The men engaged in the work believed in the project, and they needed to be able to persuade senior government and forces officials that they were worth supporting. According to Atkinson, this function was performed by Lewis.

Now that they had successfully calibrated their equipment, they continued trying to view aircraft over the next few weeks. Their observation of reflections from a Battle aircraft was noted in Dee's diary on 22 August. This was the day that Epsley and Marris from GEC came to Worth Matravers to view the 10 cm system. It was crucial that they made a good impression with a proper demonstration, and Lovell recorded in his diary that '[Epsely and Marris] very much sobered down by Hodgkin's presentation apparatus. Then in the afternoon we managed to follow a Battle for 2 miles tail on which was magnificent.'[85] This demonstration provided AMRE with enough ammunition to take control of the centimetre AI project. At the subsequent CVD meeting Dee related their success, and GEC was ordered to cancel its 25 cm work and to support AMRE's efforts.

The final major component for AMRE's apparatus to be completed was the reflecting klystron, designed by Sutton of the Admiralty valve workshop at Bristol. It arrived at the end of August. Ward and Batt had struggled to operate the 20 cm local oscillator they were using for their receiver experiments, and Sutton's valve was 'sheer magic' by comparison. Like the magnetron, it behaved in a much more stable fashion than its predecessor.[86]

CONCLUSIONS: WHAT WAS ACHIEVED

In this paper I have described the assembly of a team in the spring of 1940 to build 10 cm radar for the RAF. I have tried to show that, faced with what seemed at best an extremely difficult task, the team resorted to the kind of working relationships and methods with which they were most comfortable: those of their university laboratories. I have also tried to use the contemporary accounts to fill in the details of the characters involved in the project.

The project was rightly judged by its participants to be a success. They built a 10 cm radar under intense pressure, using scant resources, in a remarkably quick time. Despite the post-event judgements on the smoothness of their progress, when one examines original accounts written at the time, a different picture emerges. The team had had some success in managing to produce convincing demonstrations *for others.* They had also learned to produce results with their equipment. However, they were not in the position of fully understanding their apparatus and being able to replicate the equipment and the results. The participants have reconstructed the uncertain, contingent nature of their apparatus and results of this period, into something far more stable than it actually was at the time.[87] It required nearly a year's further work before 10 cm AI was stable enough to be turned into anything approaching a production apparatus.

In conclusion, the main points that I have made in this paper are as follows. First, a small close-knit team can produce very quick solutions to a problem through practical competence and close relationships. Second, the later judgement of the team regarding the smoothness of their progress can be shown to be much more contingent when viewed with the hindsight of time. When one realizes how complicated the apparatus was, one can only be even more impressed with the successful outcome.

Notes and References

1. E. Bowen, *Radar Days* (Bristol, 1987), Chapter 8.
2. A. C. B. Lovell, *Echoes of War* (Bristol, 1991), Chapter 4.
3. D. Pritchard, *The Radar War* (Wellingborough, 1989), Chapter 1.
4. Wood and Dempster, *The Narrow Margin* (London, 1961).
5. Radar was originally designed to prevent the RAF having to mount standing patrols. With radar, fighters could be sent to where they were needed, allowing the pilots to rest before they were placed directly in the way of incoming raids. This allowed successful defence with limited resources, as the Battle of Britain showed. See Wood and Dempster, *op. cit.* (4).
6. van der Hulst–Lovell correspondence, A. C. B. Lovell Collection, John Rylands University Library, Manchester University, 10 May 1940, quoted 7 February 1982.
7. J. Atkinson, 'Work at Worth Matravers and TRE Malvern', in W.E. Burcham (ed.), *Fifty Years of the Cavity Magnetron: Proceedings of a One-day Symposium* (Birmingham, 1990), 25.

8. R. Batt, *The Radar Army* (London, 1991), 15–16.

9. Atkinson, *op. cit.* (7), 25.

10. See, for example, S. Shapin, 'The Invisible Technician', *American Scientist*, 1989, 77: 754–63.

11. Batt, *op. cit.* (8), 18.

12. A. Hodgkin, *Chance and Design: Reminiscences of Science in Peace and War* (Cambridge, 1992)

13. P. Dee, Diary, in Lovell, *TRE Record* (John Rylands University Library, Manchester, 1945), 15 May 1940.

14. Hodgkin, *op. cit.* (12), 157–8. He learnt about airborne radar from Hanbury-Brown, who was probably the most experienced person in the country in terms of flying with radar equipment.

15. Batt, *op. cit.* (8), 18.

16. *Ibid.*, 33.

17. *Ibid.*, 18.

18. W. E. Burcham and E. D. R. Shearman, *Fifty Years of the Cavity Magnetron* (Birmingham, 1990), 20.

19. It is difficult to judge who was right in this particular case. Bowen was one of the original radar pioneers, and never readily accepted the appointment of Rowe, a civil servant and an armaments expert, to be head of AMRE. The distrust worsened when Rowe appointed Lewis direct from the Cavendish Laboratory to be his deputy in August 1939. This was done over the head of people like Bowen. Furthermore, Bowen's 1.5 m AI was not producing results in the summer of 1940, partly because of technical reasons and partly because of organizational ones. Nevertheless, the blame was laid at Bowen's door by Rowe hence his alienation from the others at AMRE.

20. Dee, *op. cit.* (13), 15 May 1940.

21. Interview of Burcham by author, 15 February 1993.

22. Atkinson, *op. cit.* (7), 25.

23. Dee, *op. cit.* (13), 15 May 1940.

24. Atkinson, *op. cit.* (7), 25.

25. *Ibid.*

26. Burcham, *op. cit.* (21).

27. W. E. Burcham, Laboratory Notebook, 16 July 1940.

28. Batt, *op. cit.* (8), 47.

29. Dee, *op. cit.* (13), 15 May 1940.

30. Burcham, *op. cit.* (21).

31. van der Hurst–Lovell, *op. cit.* (6), 23 May 1940.

32. Dee, *op. cit.* (13), 31 May 1940.

33. van der Hulst–Lovell, *op. cit.* (6), Lovell paper.

34. Dee, *op. cit.* (13), 5 June 1940.

35. *Ibid.*

36. *Ibid.*

37. van der Hulst–Lovell, *op. cit.* (6), letter from Dee to vdH, quoted in letter from vdH to Lovell, 7 February 1982.

38. *Ibid.*, Lovell paper.

39. *Ibid.*, Lovell diary, 4 June 1940.

40. Hodgkin, *op. cit.* (12), 161.

41. Dee, *op. cit.* (13), 10 June 1940.

42. Batt, *op. cit.* (8), 40.

43. van der Hulst–Lovell, *op. cit.* (6), Lovell diary, 12 June 1940.

44. Batt, *op. cit.* (8), 41.

45. D. C. Gooding, *Experiment and the Making of Meaning* (Dordrecht, 1990), Chapter 4.

46. Burcham, *op. cit.* (27), 16 July 1940.

47. Dee, *op. cit.* (13), 13 June 1940.

48. Burcham, *op. cit.* (27), 15 July 1940.

49. Dee, *op. cit.* (13), 14 June 1940.

50. Batt, *op. cit.* (8), 43.

51. Hodgkin, *op. cit.* (12), 161–2.

52. Dee, *op. cit.* (13), 20 June 1940.
53. Hodgkin, *op. cit.* (12), 170.
54. *Ibid.*
55. Burcham, *op. cit.* (21).
56. Atkinson, *op. cit.* (7), 26.
57. Dee, *op. cit.* (13), 9 July 1940.
58. *Ibid.*, 10 July 1940.
59. van der Hulst–Lovell, *op. cit.* (6), letter from Dee to vdH, quoted in letter from vdH to Lovell, 7 February 1982.
60. Hodgkin, *op. cit.* (12), 171.
61. Burcham, *op. cit.* (27), 12 July 1940.
62. This was the name of a ground mapping radar system developed in 1942 and captured by the Germans in 1943. They copied it in order to build their own centimetre radar.
63. Dee, *op. cit.* (13), 15 July 1940.
64. Burcham, *op. cit.* (21).
65. Burcham, *op. cit.* (27), 18 to 21 July 1940.
66. Batt, *op. cit.* (8), 50.
67. Compare to Gooding's account of how Faraday learned to operate the rotation motor by trial and error experimentation. See Gooding, *op. cit.* (45), Chapter 6.
68. Megaw was able to replicate the cavity magnetron as he had learned how to build and operate it in the first place. Part of his learning process had involved him making a trip to Birmingham to meet Randall and Boot.
69. Batt, *op. cit.* (8), 51.
70. Burcham, *op. cit.* (27), 2 August 1940.
71. van der Hulst–Lovell, *op. cit.* (6), Lovell diary, 22 July 1940.
72. Dee, *op. cit.* (13), 8 August 1940.
73. Lovell, *op. cit.* (2), 41.
74. Burcham, *op. cit.* (27), 10 August 1940.
75. Paterson, in R. J. Clayton and J. Agar, *A Scientist's War: The War Diary of Sir Clifford Paterson 1939–45* (London, 1991), 12 August 1940.
76. Burcham, *op. cit.* (27), 13 August 1940.
77. Burcham–Lovell correspondence, Lovell to Burcham, 26 February 1990.
78. Batt, *op. cit.* (8), 54.
79. *Ibid.*
80. *Ibid.*, 55.
81. Burcham, *op. cit.* (27), 12 August 1940.
82. See D. C. Gooding, 'Situating a Twentieth Century Scientist', *EASST Newsletter*, 1993, 12: 3, for further discussion of how scientists reinterpret their original notes at a later date to give the impression that they had a much clearer idea of what they were doing than is apparent from the early notes.
83. Burcham, *op. cit.* (27), 13 August 1940.
84. Atkinson, *op. cit.* (7), 26–7.
85. Lovell, *op. cit.* (2), 45.
86. Batt, *op. cit.* (8), 61.
87. For an account of the reconstructive processes that scientists use to make their apparatus and results stable, see Gooding, *op. cit.* (45), Chapter 1.

Diffusion of Brewing Technology since 1900: Change and the Consumer

TERRY GOURVISH

BREWING TECHNOLOGY AND HISTORY

The centenary of the death of Louis Pasteur (1822–95), author of *Etudes sur la bière* (1876) and the inventor of microbiology who revolutionized our understanding of the brewing process, provides an appropriate moment to reconsider the question of the relationship between the diffusion of scientific and technological innovation and the beer we drink as consumers. In brewing, as in the rest of the food and drink sector, there is often a tension between the consumer and the science-led producer. The consumer has tastes and notions which are not always logical; he or she sometimes nurses a nostalgic love of tradition, but at other times exhibits an obsession with novelty and trendiness. The producer, on the other hand, is concerned with harnessing economic benefits from science and technology, in the form of lower production costs, reliability, safety and so on. This chapter examines the nature of the brewing process, considers the way in which new techniques have been modified and translated into improved beer products (diffusion) and assesses the impact of this diffusion on the perceptions of the average British beer drinker. Above all, it is concerned with the fate of the Great British Pint, the 0.5628 litres of warm, flat ale, which with cricket, cucumber sandwiches and vicarage lawns make up some of the icons associated with Britain's pre-eminence as the world's first industrial nation.

BREWING: TRADITIONAL OR PROGRESSIVE?

The basic brewing process is relatively straightforward (see Figure 1). Malted grains, primarily barley, are mashed using warm water to release fermentable sugars. The solution, or wort, is separated from the grains, and the liquid is then boiled with hops. Once the wort has been cooled, it is fermented using yeast. The process has remained very much the same over the centuries, at least in the pre-packaging stage, and this causes many commentators to regard brewing as the epitome of conservatism and tradition. However, scientific advances and incremental improvements in brewing technology have been more important than is sometimes acknowledged. Beginning with the revolution in our understanding of fermentation associated with Pasteur and Hansen in the later nineteenth century, there have been important advances in all stages of the brewing process. Traditional methods – cask-conditioned and bottle-conditioned beers, fermented in open squares, with many other vessels open to the atmosphere and liquids moved mainly by gravity – were challenged by developments in pasteurization and filtration from the late nineteenth century. For example, as early as 1893 a patent of Fallow & Jackson provided for gas dispense using carbonic acid.[1] The long-established emphasis on the need for good quality in raw materials, absolute cleanliness in the brewing process and the search for consistency in the product encouraged technological innovation, though the pace of change was not even.

Technological change accelerated after 1960. It focused on capturing the scale economies demanded by an industry characterized by a concentrated, oligopolistic structure and translating these gains into the more effective, speedier distribution of more consistent, longer-life, mass-produced beers. The main developments were: more effective and automated control of malting, brewing, fermentation and conditioning, with the use of closed vessels, computer-aided control etc.; the development of 'continuous brewing' techniques, particularly in fermentation, and 'high-gravity' brewing; improved packaging, particularly in bottling and canning, with much faster process throughput; and more effective distribution by road.[2]

The search for consistent production is not a new phenomenon, however. We should note in particular the move Whitbread made in 1927 to replace its bottle-conditioned ale with brewery conditioned (filtered, pasteurized and carbonated) 'bright beer', which heralded a protracted move away from bottle-conditioned beer, culminating in the decision by Guinness to adopt brewery conditioning for its bottled Extra Stout.[3]

THEORIES OF INNOVATION AND TECHNOLOGICAL CHANGE

Before we examine the extent to which the improvements in consistency, safety, reliability, shelf-life and so on have been welcomed by the consumer, it is important to examine some of the main developments in brewing since 1900, and to see how they fit into three ways of regarding technological change. Historians of technology have many ways of looking at the diffusion process.[4] Here, I discuss three 'approaches' or 'theories': leading firms and first-mover advantage; path dependency and contingency; and convergence.

Leading Firms and First-mover Advantage

The concept of first-mover advantage exploited by leading and often large-scale firms figures prominently in the business history literature.[5] At first sight, the model fits the brewing experience very well. Brewers such

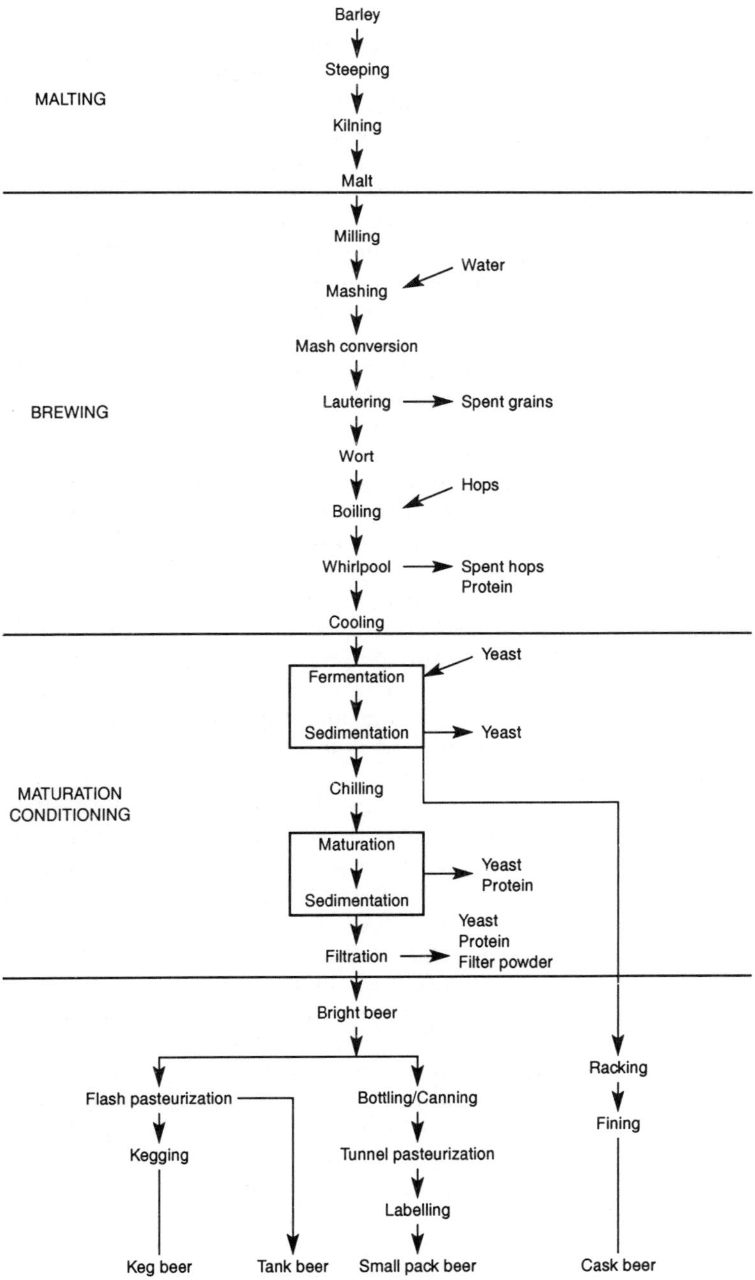

Figure 1 The brewing process.
Source: Brewers and Licensed Retailers Association.

as Bass and Allsopp led the way in employing chemists and establishing laboratories in the second half of the nineteenth century, as did Guinness, which employed the brilliant chemist Horace Tabberer Brown as director of its research laboratory in Dublin. Lacking a large estate of tied public houses in the UK, Guinness continued to place considerable emphasis on research and development in the twentieth century. London brewers such as Watney, Truman and Whitbread were also prominent in the research field after 1900. These firms clearly come into the category of leading firms using a resource-based strategy to accumulate valuable technology assets.[6] They may be said to have created a competitive advantage with their new products, Bass and Allsopp in pale ale, Guinness in stout, Whitbread in pasteurized pale ale, Watney in keg bitter. However, these beer styles could be (and were) copied by other brewers, and the evidence suggests that much change in brewing took the form of a 'learning curve', i.e. incremental improvements to basic processes and materials. The progress that was made tended to be gradual, the result of continuous adjustments to brewing techniques in the search for quality, consistency and economy, rather than via dramatic scientific discoveries emerging from the technological leadership of a leading edge firm, with new processes protected by patents.[7]

Path Dependency and Contingency

The popularity of the 'social constructionist' approach to technology challenges the notion of scientific determinism and emphasizes instead the importance of social processes.[8] Path dependency, in which particular technologies emerge and retain their advantage, was the subject of much debate when Paul David sought to explain the persistence of the QWERTY configuration on typewriter/processor keyboards.[9] In brewing, path dependency may be seen in the adherence to 'top fermentation' techniques and in the survival of the beer engine as a means of moving beer from the cellar to the public house bar. The retention of the non-metric pint and the preference in the UK for drinking outside the home have also been important constraints on the possibilities for technological change, especially in packaging.

Contingency, or luck, is another element to find favour with modern analysts.[10] Certainly, it may have been a matter of luck that the new lager type of beer failed to take off in the UK in the 1880s, as the trade expected, following its popularity in continental Europe and subsequently in the USA and elsewhere. In the UK, the fortuitous switch to pale ales of the Burton type may have satisfied the market, making further innovation unattractive. Conversely, lager's successful assault on the British market from 1960 owed something at least to a run of hot summers, and the reluctance or slowness of major brewers to improve the cellar cooling of bitter beers in order to preserve their condition in periods of hot weather. There was also some irony (given the experience of the 1880s) in the fact that a technological change in the *bitter* market (keg beer) probably helped to prepare the ground for a shift to draught (keg) lager by consumers.[11]

Convergence

Convergence of technological processes is another element worthy of examination. Historians have identified a tendency for technologies to convergence over time, and for certain technologies to become fixed or 'closed'.[12] In brewing, convergence was assisted by the degree of co-operation and information sharing which took place at the corporate and industry level. Brewing companies were certainly secretive about their production figures and retail outlets, and Guinness was notable for its reluctance to share the results of its concentrated work on barley and hops in the inter-war years.[13]

However, the evidence suggests that from the 1950s there was much more inter-firm co-operation with scientific and technical information, encouraged by national bodies such as the Institute of Brewing, the Incorporated Brewers' Guild and the Brewers' Society (now Brewers and Licensed Retailers Association), and expressed via the work of the Brewing (Industry) Research Foundation. The foundation, established in 1951 on the joint initiative of the Institute of Brewing and the Brewers' Society, was led first by Professor Sir Ian Heilbron of Imperial College and then by Dr Arthur Cook. Co-operating with the larger companies in joint projects, it undertook important research into all aspects of brewing, including the hitherto neglected areas of wort production and fermentation, and altogether spent £20 million in 1980 prices over the period 1955–80.[14]

The co-operative environment was encouraged by the incremental nature of change. Thus, in the later twentieth century, research continues, as it has done for 150 years, on developing improved barley and hop varieties, and controlling the brewing and fermenting processes. Research on 'continuous brewing', a continuous flow process aiming to secure scale economies and significant savings in fuel, was the subject of experimental work by the Brewing Industry Research Foundation from 1956, and had been translated into commercial production by Bass and Watney Mann & Truman by the early 1970s. Although the results were disappointing in terms of costs, the experience gained was applied to improvements obtained from 'accelerated batch' and 'high gravity' brewing techniques in the 1970s, and work is still proceeding in these areas.[15] With raw materials attention is focused currently, for example, on identifying pro-foaming agents and reducing pesticide residues in malt and hops. Incremental improvements in milling, mashing, wort separation and boiling, and in the automatic cleaning of vessels, are being made. In packaging and dispense there have been important developments with in-can systems (draught beer in a can) and with the use of nitrogen (N_2) in mixed-gas systems (see below).[16]

SCIENCE, NEW BEERS AND THE CONSUMER

Beer products, like their markets, are currently in a state of flux, encouraged by changes in the regulatory regime facing brewing, notably the British government's Beer Orders of 1989, the challenge of freer

Table 1 *UK beer market trends*

Year	Lager as percentage of market	Canned beer as percentage of market	Take-home trade as percentage of market
1960	1	1	5
1974	16	6	10
1993	52	23	25

Source: Brewers and Licensed Retailers Association, *Statistical Handbook 1994*.

European markets and recession. The Beer Orders have had a particular effect, attacking the integrated brewing–retailing operations of the larger brewers, such that the number of brewery-tied public houses has fallen from 44,000 (54 per cent of the total) in 1989 to 30,000 (36 per cent) in 1994.[17] In this current state of market uncertainty, the interesting question to pose is whether the new beers produced as a result of technical change are leading or following the market.

What has happened? First, new technology has helped to encourage a degree of product convergence. There has been an increasing preference for lighter, sweeter, blander beers. Second, the temperature at which beer is served has also fallen, as is evident not only with lager, which is usually kegged, pasteurized, filtered, and gas-dispensed, but also with ales and stouts. Third, there has been a move to more drinking at home and to packaging in cans (see Table 1). On the other hand, cask-conditioned ale remains popular. It is a substantial niche market which has increased its market share in the 1990s. Real ale sales rose by 9 per cent in 1993, while keg beer declined by 4 per cent; real ale made up 44 per cent of the draught bitter/stout market in 1993 compared with 34 per cent in 1989, and had a share of the total market of about 17 per cent.[18]

Cause and effect are difficult to disentangle, of course. Changing consumer tastes and habits undoubtedly had an impact on the demand for beer. The following elements were clearly important: the emergence of segmented markets, and especially the 'youth market'; the experiences acquired from foreign holidays; a greater health consciousness, which often expressed itself in preferences for lighter coloured drinks; changing shopping and leisure patterns, and in particular the increasing use of the home as an 'entertainment centre'. On the other hand, technology itself can be a potent force for product change, and the aggressive marketing of Watney's Red Barrel in the 1960s is merely one example of the way in which major brewing companies succeeded in 'leading' consumers. A better known case from the United States was the spectacular success of Philip Morris-Miller's launch of 'Miller Lite' in 1975.[19]

Those who point to brewing companies as the driving force for product change customarily see consumer groups such as CAMRA, the

Campaign for Real Ale set up in 1971 to lobby for traditional beer, as rather weak players in the game. For example, when MORI examined consumer awareness in 1975, it found that only 12 per cent of those interviewed had heard of CAMRA. However hard CAMRA protested about 'unreal' beers such as keg bitter and lager and beer in cans, the strength of the shift in the beer market revealed in Table 1 could not have occurred without the support and even enthusiasm of a substantial section of beer drinkers.[20] On the other hand, CAMRA was the focal point for the assertion of consumer preferences which should not be underestimated. The spectacular failure of Watney's 'Red' and 'Red Revolution' campaigns in 1971–4 are a clear testimony to this.[21] Consumer passivity, given the gradual nature of change, should also not be ruled out.

To what extent has the industry encouraged the shifts in beer drinking – the move to lager, drinking at home and canned beer? Recent events reveal a similar mix of technologically driven change by companies and conflicting, often equivocal, consumer attitudes. Here I look in particular at the new 'widget' and draught 'mixed gas' beers.

The Inexorable March of the Widget: Post Widget Reductionism?

The appearance in 1988 of the 'widget' has been hailed as one of the most important technical developments in British brewing for many years. Following Guinness's pioneering work on the concept of replicating draught beer in a can, which was applied first to canned Draught Guinness, then to a new Guinness Draught Bitter,[22] at least six separate types of in-can system (ICS) emerged. These are designed to provide canned beer with low carbonization and a creamy head when poured. Widgets are usually moulded plastic inserts which supply or nucleate CO_2 and/or N_2 bubbles. Guinness, Whitbread (Draughtflow, 1991, and an in-bottle system, 1995) and Bass (Super-widget system, 1994) in particular have exploited this new technology.[23] The most important brands in this small but rapidly growing segment of the beer market are Boddingtons, which enjoyed a 60 per cent market share of ICS beers in 1993/4, Worthington, Guinness, John Smith and Theakston.[24]

Creamy Draught Beers dispensed with Mixed Gas

Mixed gas systems (CO_2 and N_2) for draught beers have also emerged. They achieve reduced waste, improved cost-effectiveness, faster speed of dispense and a tighter, foamier head, on top of the original advantages of keg beer – a sterile environment, free of beer-spoiling organisms.[25] Guinness was also the pioneer here. Its innovative Draught Guinness, launched in 1961, was delivered with the help of nitrogen and 'Easy Serve' casks.[26] Recently, there has been some evidence of technological convergence, in that certain draught beers, mainly keg, have aped some of the ICS developments to produce a less gassy beer with a creamier foaming head. This has been achieved by means of mixed gas dispense (notably with Caffreys), or through the use of special beer-pump fittings, i.e. thin, swan-necked founts and tighter sparklers, which force the beer through narrower apertures, thus producing an agitation in the beer which gives a highly aerated pint settling to a rich creamy head (e.g. Tetleys).

Is it possible to state that the consumer has welcomed these developments? Market research has certainly identified a switch to ICS beers such as Boddingtons, but it is not clear how far this is a matter of choice, and how far a matter of supermarket purchasing policies. Cooler, creamier-head beers are certainly different from the serving of traditional, cellar-cooled bitters by companies such as Adnams. It is also too soon to assess whether the success of Caffreys has been as much advertising-led as owing to public awareness of the technical advance in this keg beer.

There are complications to a simplified picture, of course. First, there is the issue of 'blanket pressure' or 'cask-breather' ale. Here, gas (usually CO_2) is used to exclude ambient air in a cask after opening. Does this disqualify beer handled in this way from being called 'real ale'? CAMRA dismisses the practice; but the trade insists that there is no evidence for an increase in carbonization as a result.[27] Second, some brewers have taken steps to reduce the level of secondary fermentation in their cask beers. Again, does this prevent them from enjoying the appellation 'real ale'? The beer has not been pasteurized, but is not the 'living' beer praised by CAMRA. We should also note that technology can serve small, 'micro' brewers too, with special-purpose smaller and more flexible equipment, allowing production to respond economically to changes in taste or product mix.[28] Reunion Cellar Equipment Ltd has developed a cellar 'widget' for cask beer, to ensure that beer is drawn from the top of the cask.[29] Finally, product development and demand shifts can take unexpected turns. Some analysts have observed that the development of the ICS system has also had the effect of encouraging drinkers to return to the pub to purchase draught beer.[30]

Consumers do not always appreciate technical progress. Thus, when a test was made in the Northampton area in the 1960s of a beer produced by continuous wort boiling, it was roundly condemned for tasting 'like nail varnish'.[31] Brewers, who have done much to embrace technological innovation since 1950, have always had to strike a balance between science and tradition, between lower operating costs and higher capital costs. To understand the process fully we have to fuse two distinct theoretical approaches to technological change: that which emphasizes leading firms and first-mover advantage; and that which accepts the importance of social processes, tradition (which encourages path-dependency), inter-firm co-operation (which encourages convergence) and contingency (in which capricious consumer attitudes play their part). The challenge the brewing industry still faces is 'to present an acceptable product to a public which covers the whole gamut from undiscerning to pernickety: to please the public which drinks merely by eye, as much as that section which drinks by eye, nose and palate'.[32]

Consumer groups such as CAMRA, in criticizing the new products of the 1990s, are articulating long-established tensions between the tastes of some consumers and product-enhancing, cost-reducing innovation, tensions between consumer wants and the profit motives and market share ambitions of major oligopolistic beer producers. Neither is static, of course. CAMRA had considerable success in slowing down the march of keg beers in the 1970s and 1980s. Its actions have constituted a major resistance campaign on the part of committed beer drinkers and have had a major impact in widening consumer choice in pubs and in preserving the supply of traditional ales. The lobby patently caused companies such as Allied and Grand Metropolitan, owners of Watney Mann & Truman, to redirect their marketing strategies.[33] It remains to be seen whether this pressure group of beer buffs and middle-class drinkers can continue to champion the 'craft' production of real ale, with British brewing, along with its continental European partners, apparently on the threshold of a process of corporate globalization and international branding.

Notes and References

This article is a revised version of a paper entitled 'Has technology ruined the Great British Pint?', given at the Joint History of Science and Engineering Section of the annual meeting of the British Association for the Advancement of Science, Newcastle-upon-Tyne, in September 1995, and subsequently at the seminar of the South Kensington Institute for the History of Technology in October 1995. I should like to thank the organizers and participants for their helpful comments.

1. See B. Richards, 'Mixed Gas Dispense', *The Brewer*, March 1994: 109.

2. See T.R. Gourvish and R.G. Wilson, *The British Brewing Industry 1830–1980* (Cambridge, 1994), 534–48.

3. Gourvish and Wilson, *op. cit.* (2), 348, 353–4.

4. Among the burgeoning literature see J. Elster, *Explaining Technical Change: A Case Study in the Philosophy of Science* (Cambridge, 1983); Giovanni Dosi *et al.* (eds), *Technical Change and Economic Theory* (1988); T.P. Hughes, *Networks of Power: Electrification in Western Society, 1880–1930* (Baltimore, 1983); P. Paulo Saviotti and J. Stanley Metcalfe (eds), *Evolutionary Theories of Economic and Technological Change* (Chur, 1991); R.R. Nelson and S.G. Winter, 'In Search of a Useful Theory of Innovation', *Research Policy*, 1977, 6: 36–76 and *An Evolutionary Theory of Economic Change* (Cambridge, MA, 1982); D. Teece and G. Pisano, 'The Dynamic Capabilities of Firms: An Introduction', *Industrial and Corporate Change*, 1994, 3: 537–56, pp. 546–8; and D.A. Hounshell, 'Hughesian History of Technology and Chandlerian Business History: Parallels, Departures and Critics', *History and Technology*, 1995, 12: 205–24.

5. Not least by A.D. Chandler Jr, *Scale and Scope: The Dynamics of Industrial Capitalism* (Cambridge, MA, 1990), and see the literature review by M.B. Lieberman and D.B. Montgomery, 'First-mover Advantages', *Strategic Management Journal*, 1988, 9: 41–58.

6. Cf. Teece and Pissano, *op. cit.* (4).

7. Cf. Lieberman and Montgomery, *op. cit.* (5), 42–3.

8. Cf. W.E. Bijker, T.P. Hughes and T.J. Pinch (eds), *The Social Construction of Technological Systems: New Directions in the Sociology and History of Technology* (Cambridge, MA, 1987); W.E. Bijker, *Of Bicycles, Bakelites and Bulbs: Toward a Theory of Sociotechnical Change* (Cambridge, MA, 1995); Hughes, *op. cit.* (4), 212–14.

9. P. A. David, 'Clio in the Economy of QWERTY', *American Economic Review*, 1985, 75: 332–7, and 'Understanding the Economics of QWERTY: The Necessity of History', in W.N. Parker (ed.), *Economic History and the Modern Economist* (New York, 1986); for a broader summary of the issue see O.E. Williamson, 'Transaction Cost Economics and Organization Theory', *Industrial and Corporate Change*, 1993, 2: 107–56, pp. 142–3.

10. Cf. G. Carroll and R. Harrison, cited in Williamson, *op. cit.* (9), 142–3.

11. This is not to dismiss the importance of brewing companies' corporate strategies, and especially aggressive marketing and the use of production licensing agreements. See

Richard G. Wilson and T.R. Gourvish, 'The Foreign Dimensions of British Brewing (1880–1980)', in E. Aerts, L.M. Cullen and R.G. Wilson (eds), *Production, Marketing and Consumption of Alcoholic Beverages* (Leuven, 1990), 124–6, 131–3; Gourvish and Wilson, *op. cit.* (2), 457–9.

12. Cf. Bijker *et al.*, *op. cit.* (8), 4, 12–13; R.N. Langlois and P.L. Robertson, *Firms, Markets and Economic Change* (1995), 70–1.

13. Gourvish and Wilson, *op. cit.* (2), 536.

14. Gourvish and Wilson, *op. cit.* (2), 538–41; A.D. Portno, 'Brewing Research Foundation: Twenty-five Years of Brewing Research', *Journal of the Institute of Brewing*, June 1983, 89: 146–8; C. Dalgleish, '25 Years of Brewing Research', *Brewing Review*, March 1976, 109–11; and cf. D.J. Teece, 'Technological Change and the Nature of the Firm', in Dosi *et al.*, *op. cit.* (4), 275–9. On Heilbron and Cook, see *Journal of the Institute of Brewing*, 1960, 66: 10–11; 1989, 95: 11–12.

15. L.R. Bishop, 'A System of Continuous Fermentation', *Journal of the Institute of Brewing*, 1970, 76: 172–81; A.H. Button, 'Continuous and Rapid Batch Processes', *Brewers' Guardian*, July 1970: 50–7; A.D. Portno, 'Continuous Fermentation – Past, Present & Future', *Brewing Review*, April 1976: 170–5; P.J. Hawkins, 'High Gravity Brewing', *Brewers' Guardian*, May 1975: 46–7; C. Whitworth, 'High Gravity Brewing', *The Brewer*, November 1980: 339–43.

16. Cf. C. Marchbanks, *Brewing and Distilling International*, April 1990: 16–17; *idem.*, May 1995; A.J.L. Kennedy, *The Brewer*, July 1993: 292–6; E.A. Sherrington and G.D. Roberts, *ibid.*, 302–6.

17. *Brewers and Licensed Retailers Association Statistical Handbook 1994–5*; T. Gourvish, 'The impact of EC Legislation on Brewing Industry Competition', *Brewers' Guardian*, August 1995, 124: 31–4, pp. 33–4.

18. *Brewing and Distilling International*, October 1994: 4.

19. W.J. Reader and J. Slinn (of Oxford Brookes University), unpublished history of Grand Metropolitan, Chapter 8; K.G. Elzinga, 'The Beer Industry', in W. Adams (ed.), *The Structure of American Industry* (New York, 6th edn, 1982), 218–48, pp. 234–6; J. Sutton, *Sunk Costs and Market Structure: Price Competition, Advertising, and the Evolution of Concentration* (Cambridge, MA, 1992), 295–6; M.E. Porter, *Competitive Strategy: Techniques for Analyzing Industries and Competitors* (New York, 1980), 60. There was, of course, a technological element in Miller Lite, the use of amylglucosidase, a carbohydrate-reducing enzyme.

20. MORI, 'Attitudes to the British Brewing Industry' (Brewers and Licensed Retailers Association, 1975), vii.

21. W. Greenwell *et al.*, *The Brewing Industry. Quarterly Commentary*, August 1976: 14.

22. Note the work of Alan Forage and William Byrne at Guinness in 1984, which led to the 1987 patent. See Keynote, *Breweries and the Beer Market*, 1994: 52; David Newton, Q&A, in *New Scientist*, 9 September 1995.

23. *Brewing and Distilling International*, September 1994 and May 1995; *The Brewer*, May 1995. Bass's 'super-widget' is made of aluminium and cost £10 million to develop. It obviates the need to refrigerate the can to prevent fobbing when poured. However, Whitbread makes the same claim for its 'Draughtflow' system. Whitbread's new in-bottle system was developed at PLM Redfearn's glass packaging plant in Barnsley.

24. Keynote, *op. cit.* (18), 53; Taylor Nelson AGB LiquorPanel data, information from Nick Battinson. Note that some beers regarded by analysts as 'widgeted' are bitters canned under nitrogen without a widget device.

25. Cf. R. Zamanek, *The Brewer*, November 1992; L. Wilcox, *The Brewer*, March 1994.

26. J.R. Moore, 'History of Draught Guinness', typescript, 1974, Guinness plc archive, Park Royal, London.

27. Cf. editorial in *The Brewer*, February 1995.

28. J. Riley, 'Brewing Small', *The Brewer*, May 1995: 180–4.

29. *Brewing and Distilling International*, September 1994.

30. Cf. Liz Morgan of Carlsberg-Tetley, quoted in *Brewing & Distilling International*, October 1994.

31. Gourvish and Wilson, *op. cit.* (2), 548.

32. Prof. I. Preece, 1964, cited in *ibid.*, 548.

33. R. Protz and T. Millns (eds), *Called to the Bar: An Account of the First 21 Years of the Campaign for Real Ale* (St Albans, 1992).

Contents of Former Volumes

FIRST ANNUAL VOLUME, 1976*

D. S. L. CARDWELL and RICHARD L. HILLS, Thermodynamics and Practical Engineering in the Nineteenth Century.
JACQUES HEYMAN, Couplet's Engineering Memoirs, 1726–33.
NORMAN A. F. SMITH, Attitudes to Roman Engineering and the Question of the Inverted Siphon.
R. A. BUCHANAN, The Promethean Revolution: Science, Technology and History.
M. DAUMAS, The History of Technology: Its Aims, its Limits, its Methods.
KEITH DAWSON, Electromagnetic Telegraphy: Early Ideas, Proposals and Apparatus.
MARIE BOAS HALL, The Strange Case of Aluminium.
G. HOLLISTER-SHORT, Leads and Lags in Late Seventeenth-Century English Technology.

SECOND ANNUAL VOLUME, 1977*

EMORY L. KEMP, Samuel Brown: Britain's Pioneer Suspension Bridge Builder.
DONALD R. HILL, The Banū Mūsá and their 'Book of Ingenious Devices'.
J. F. CAVE, A Note on Roman Metal Turning.
J. A. GARCIA-DIEGO, Old Dams in Extremadura.
G. HOLLISTER-SHORT, The Vocabulary of Technology.
RICHARD L. HILLS, Museums, History and Working Machines.
DENIS SMITH, The Use of Models in Nineteenth-Century British Suspension Bridge Design.
NORMAN A. F. SMITH, The Origins of the Water Turbine and the Invention of its Name.

THIRD ANNUAL VOLUME, 1978*

JACK SIMMONS, Technology in History.
R. A. BUCHANAN, History of Technology in the Teaching of History.
P. B. MORICE, The Role of History in a Civil Engineering Course.
JOYCE BROWN, Sir Proby Cautley (1802–71), a Pioneer of Indian Irrigation.
A. RUPERT HALL, On Knowing, and Knowing How to . . .
FRANK D. PRAGER, Vitruvius and the Elevated Aqueducts.
JAMES A. RUFFNER, Two Problems in Fuel Technology.
JOHN C. SCOTT, The Historical Development of Theories of Wave-Calming Oil.

FOURTH ANNUAL VOLUME, 1979*

P. S. BARDELL, Some Aspects of the History of Journal Bearings and their Lubrication.
K. R. FAIRCLOUCH, The Waltham Pound Lock.
ROBERT FRIEDEL, Parkesine and Celluloid: The Failure and Success of the First Modern Plastic.
J. G. JAMES, Iron Arched Bridge Designs in Pre-Revolutionary France.
L. J. JONES, The Early History of Mechanical Harvesting.
G. HOLLISTER-SHORT, The Sector and Chain: An Historical Enquiry.

FIFTH ANNUAL VOLUME, 1980*

THOMAS P. HUGHES, The Order of the Technological World.
THORKILD SCHIØLER, Bronze Roman Piston Pumps.
STILLMAN DRAKE, Measurement in Galileo's Science.
L. J. JONES, John Ridley and the South Australian 'Stripper'.
D. G. TUCKER, Emile Lamm's Self-Propelled Tramcars 1870–72 and the Evolution of the Fireless Locomotive.
S. R. BROADBRIDGE, British Industry in 1767: Extracts from a Travel Journal of Joseph Banks.

RICHARD L. HILLS, Water, Stampers and Paper in the Auvergne: A Medieval Tradition.

SIXTH ANNUAL VOLUME, 1981*

MARJORIE NICE BOYER, Moving Ahead with the Fifteenth Century: New Ideas in Bridge Construction at Orleans.
ANDRÉ WEGENER SLEESWYK, Hand-Cranking in Egyptian Antiquity.
CHARLES SÜSSKIND, The Invention of Computed Tomography.
RICHARD L. HILLS, Early Locomotive Building near Manchester.
L. L. COATSWORTH, B. I. KRONBERG and M. C. USSELMAN, The Artefact as Historical Document. Part 1: The Fine Platinum Wires of W. H. Wollaston.
A. RUPERT HALL and N. C. RUSSELL, What about the Fulling-Mill?
MICHAEL FORES, *Technik*: Or Mumford Reconsidered.

SEVENTH ANNUAL VOLUME, 1982*

MARJORIE NICE BOYER, Water Mills: A Problem for the Bridges and Boats of Medieval France.
Wm. DAVID COMPTON, Internal-Combustion Engines and their Fuel: A Preliminary Exploration of Technological Interplay.
F. T. EVANS, Wood since the Industrial Revolution: A Strategic Retreat?
MICHAEL FORES, Francis Bacon and the Myth of Industrial Science.
D. G. TUCKER, The Purpose and Principles of Research in an Electrical Manufacturing Business of Moderate Size, as Stated by J. A. Crabtree in 1930.
ROMAN MALINOWSKI, Ancient, Mortars and Concretes: Aspects of their Durability.
V. FOLEY, W. SOEDEL, J. TURNER and B. WILHOITE, The Origin of Gearing.

EIGHT ANNUAL VOLUME, 1983*

W. ADDIS, A New Approach to the History of Structural Engineering.
HANS-JOACHIM BRAUN, The National Association of German-American Technologists and Technology Transfer between Germany and the United States, 1884–1930.
W. BERNARD CARLSON, Edison in the Mountains: The Magnetic Ore Separation Venture, 1879–1900.
THOMAS DAY, Samuel Brown: His Influence on the Design of Suspension Bridges.
ROBERT H. J. SELLIN, The Large Roman Water Mill at Barbegal (France).
G. HOLLISTER-SHORT, The Use of Gunpowder in Mining: A Document of 1627.
MIKULÁŠ TEICH, Fermentation Theory and Practice: The Beginnings of Pure Yeast Cultivation and English Brewing, 1883–1913.
GEORGE TIMMONS, Education and Technology in the Industrial Revolution.

NINTH ANNUAL VOLUME, 1984*

P. S. BARDELL, The Origins of Alloy Steels.
MARJORIE NICE BOYER, A Fourteenth-Century Pile Driver: The *Engin* of the Bridge at Orléans.
MICHAEL DUFFY, Rail Stresses, Impact Loading and Steam Locomotive Design.
JOSÉ A. GARCIA-DIEGO, Giovanni Francesco Sitoni, an Hydraulic Engineer of the Renaissance.
DONALD R. HILL, Information on Engineering in the Works of Muslim Geographers.
ROBERT J. SPAIN, The Second-Century Romano-British Watermill at Ickham, Kent.
IAN R. WINSHIP, The Gas Engine in British Agriculture, *c.* 1870–1925.

TENTH ANNUAL VOLUME, 1985*

D. de COGAN, Dr E. O. W. Whitehouse and the 1858 trans-Atlantic Cable.
A. RUPERT HALL, Isaac Newton's Steamer.
G. J. HOLLISTER-SHORT, Gunpowder and Mining in Sixteenth- and Seventeenth-Century Europe.
G. J. JACKSON, Evidence of American Influence on the Designs of Nineteenth-Century Drilling Tools, Obtained from British Patent Specifications and Other Sources.
JACQUES PAYEN, Beau de Rochas Devant la Technique et l'Industrie de son Temps.

ÖRJAN WIKANDER, Archaeological Evidence for Early Water-Mills – An Interim Report.
A. P. WOOLRICH, John Farey and the Smeaton Manuscripts.
MIKE CHRIMES, Bridges: A Bibliography of Articles Published in Scientific Periodicals 1800–1829.

ELEVENTH ANNUAL VOLUME, 1986*

HANS-JOACHIM BRAUN, Technology Transfer under Conditions of War. German Aero-Technology in Japan during the Second World War.
VERNARD FOLEY, with SUSAN CANGANELLI, JOHN CONNOR and DAVID RADER, Using the Early Slide-Rest.
J. G. JAMES, The Origins and Worldwide Spread of Warren-Truss Bridges in the Mid-Nineteenth Century. Part 1: Origins and Early Examples in the UK.
ANDREW NAHUM, The Rotary Aero Engine.
DALE. H. PORTER, An Historian's Judgments about the Thames Embankment.
JOHN H. WHITE, More Than an Idea Whose Time Has Come: The Beginnings of Steel Freight Cars.
IAN R. WINSHIP, The Acceptance of Continuous Brakes on Railways in Britain.

TWELFTH ANNUAL VOLUME, 1990

KENNETH C. BARRACLOUGH, Swedish Iron and Sheffield Steel.
IAN INKSTER, Intellectual Dependency and the Sources of Invention: Britain and the Australian Technological System in the Nineteenth Century.
M. T. WRIGHT, Rational and Irrational Reconstruction: The London Sundial-Calendar and the Early History of Geared Mechanisms.
J. V. FIELD, Some Roman and Byzantine Portable Sundials and the London Sundial-Calendar.
R. T. McCUTCHEON, Modern Construction Technology in Low-Income Housing Policy: The Case of Industrialized Building and the Manifold Links between Technology and Society in an Established Industry.
Book Review by Frank A. J. L. James: André Guillerme, *Le Temps de l'eau: la cité, l'eau et les techniques: nord de la France fin IIIe–début XIXe siècle.* Eng. trans.: *The Age of Water: The Urban Environment in the North of France,* AD 300–1800.

THIRTEENTH ANNUAL VOLUME, 1991*

BRUCE, J. HUNT, Michael Faraday, Cable Telegraphy and the Rise of Field Theory.
IWAN R. MORUS, Telegraphy and the Technology of Display: The Electricians and Samuel Morse.
BRIAN GEE, Electromagnetic Engines: Pre-Technology and Development Immediately Following Faraday's Discovery of Electromagnetic Rotations.
GRAEME GOODAY, Teaching Telegraphy and Electrotechnics in the Physics Laboratory: William Ayrton and the Creation of an Academic Space for Electrical Engineering in Britain 1873–1884.
W. J. READER, 'The Engineer Must Be a Scientific Man': The Origins of the Society of Telegraph Engineers.
C. A. HEMPSTEAD, An Appraisal of Fleeming Jenkin (1833–1885), Electrical Engineer.
A. C. LYNCH, The Sources for a Biography of Oliver Heaviside.
W. BERNARD CARLSON, Building Thomas Edison's Laboratory at West Orange, New Jersey: A Case Study in Using Craft Knowledge for Technological Invention, 1886–1888.
BRIAN BOWERS, Edison and Early Electrical Engineering in Britain.
R. W. BURNS, The Contributions of the Bell Telephone Laboratories to the Early Development of Television.
JONATHAN COOPERSMITH, Technology Transfer in Russian Electrification, 1870–1925.
G. HOLLISTER-SHORT, ICOHTEC XVIII Conference Report: A Personal View.

FOURTEENTH ANNUAL VOLUME, 1992

HANS-JOACHIM BRAUN, Aero-Engine Production in the Third Reich.
ANDRÉ GUILLERME, Chaleur et Chauffage: L'Introduction du Confort à Paris sous la Restauration (English summary by G.H-S).
JOHN LANGDON, The Birth and Demise of a Medieval Windmill.

MICHAEL J. T. LEWIS, The South-Pointing Chariot in Rome: Gearing in China and the West.
STANLEY SMITH, The Development and Use of the Tubular Beam, 1830–1860.
NICOLÁS GARCÍA TAPIA, Some Designs of Jerónimo de Ayanz (*c.* 1550–1613) relating to Mining, Metallurgy and Steam Pumps.
DONALD E. TARTER, Peenemünde and Los Alamos: Two Studies.
GRAHAM HOLLISTER-SHORT, Reflections on Two Conferences: ICOHTEC XIX and MASTECH.
Announcement: The Georgius Agricola Commemorations, 1994.

FIFTEENTH ANNUAL VOLUME, 1993

MICHAEL E. GORMAN, MATTHEW M. MEHALIK, W. BERNARD CARLSON AND MICHAEL OBLON, Alexander Graham Bell, Elisha Gray and the Speaking Telegraph: A Cognitive Comparison.
GRAHAM HOLLISTER-SHORT, On the Origins of the Suction Lift Pump.
ALEX KELLER, Technological Aspirations and the Motivation of Natural Philosophy in Seventeenth-Century England.
WALTER ENDREI, Count Theodore Batthyány's Paddle-Wheel Ship.
BJÖRN IVAR BERG, The Kongsberg Silver Mines and the Norwegian Mining-Museum.
HERMANN KNOFLACHER, Does the Development of Mobility in Traffic Follow a Pattern?
MICHAEL J. T. LEWIS, The Greeks and the Early Windmill.
GRAEME GOODAYE, Faraday Reinvented: Moral Imagery and Institutional Icons in Victorian Electrical Engineering.
PHILIPPE BRAUNSTEIN, Légendes Welsches et Itinéraires Silésiens: La Prospection Minière au XVe Siècle (English summary by G.H-S).

SIXTEENTH ANNUAL VOLUME, 1994

WILLIAM J. PIKE, Drilling Technology Transfer between North America and the North Sea: The Semi-Submersible Drilling Unit.
GUNNAR NERHEIM, The Condeep Concept: The Development and Breakthrough of Concrete Gravity Platforms.
W. DAVID LEWIS, Edward A. Uehling and the Automatic Pig-Casting Machine: A Case-Study of Technological Transfer.
DONALD HILL, The Toledo Water-Clocks of *c.* 1075.
GRAHAM HOLLISTER-SHORT, The Other Side of the Coin: Wood Transport Systems in Pre-Industrial Europe.
ALAN WILLIAMS, The Blast Furnace and the Mass Production of Armour Plate.
MELVIN KRANZBERG, ICOHTEC: Some Informal Personal Reminiscences.
MICHAEL FORES, Hamlet without the Prince: The Strange Death of Technical Skill in Histories.
WERNER KROKER, History of Technology at the German Mining Museum, Bochum.
R. ANGUS BUCHANAN, The Structure of Technological Revolution.
HANS-JOACHIM BRAUN, Reforming ICOHTEC: International Committee for the History of Technology.

SEVENTEENTH ANNUAL VOLUME, 1995

SUSAN MURPHY, Heron of Alexandria's *On Automaton-Making.*
D. L. SIMMS, Archimedes the Engineer.
THOMAS F. GLICK, Moriscos and Marranos as Agents of Technological Diffusion.
CARROLL PURSELL, Variations on Mass Production: The Case of Furniture Manufacture in the United States to 1940.
JENNIFER TANN, Space, Time and Innovation Characteristics: The Contribution of Diffusion Process Theory to the History of Technology.
DAVID BRIDGE, The German Miners at Keswick and the Question of Bismuth.
WALTER ENDREI, Jean Errard (1554–1610) and His Book of Machines: *Le Premier Livre des instruments mathématiques méchaniques* of 1584.
HELLMUT JANETSCHEK, From the Imperial-Royal Collection of Manufactured Products to the Museum of Technology and Industry in Vienna.
GRAHAM HOLLISTER-SHORT, Cranks and Scholars.
*Out of print.